An Astronomical Inclusion Revolution

Revolution

Advancing Diversity, Equity, and Inclusion in Professional
Astronomy and Astrophysics

Online at: https://doi.org/10.1088/2514-3433/ad2174

AAS Editor in Chief

Ethan Vishniac, Johns Hopkins University, Maryland, USA

About the program:

AAS-IOP Astronomy ebooks is the official book program of the American Astronomical Society (AAS) and aims to share in depth the most fascinating areas of astronomy, astrophysics, solar physics, and planetary science. The program includes publications in the following topics:

GALAXIES AND COSMOLOGY

INTERSTELLAR MATTER AND THE LOCAL UNIVERSE

STARS AND STELLAR PHYSICS

EDUCATION, OUTREACH, AND HERITAGE

HIGH-ENERGY PHENOMENA AND FUNDAMENTAL PHYSICS

THE SUN AND THE HELIOSPHERE

THE SOLAR SYSTEM, EXOPLANETS, AND ASTROBIOLOGY

LABORATORY ASTROPHYSICS, INSTRUMENTATION, SOFTWARE, AND DATA

Books in the program range in level from short introductory texts on fast-moving areas, graduate and upper-level undergraduate textbooks, research monographs, and practical handbooks.

For a complete list of published and forthcoming titles, please visit iopscience.org/books/aas.

About the American Astronomical Society

The American Astronomical Society (aas.org), established 1899, is the major organization of professional astronomers in North America. The membership (~7,000) also includes physicists, mathematicians, geologists, engineers, and others whose research interests lie within the broad spectrum of subjects now comprising the contemporary astronomical sciences. The mission of the Society is to enhance and share humanity's scientific understanding of the universe.

An Astronomical Inclusion Revolution

Advancing Diversity, Equity, and Inclusion in Professional
Astronomy and Astrophysics

Edited by
Dara Norman
NSFs NOIRLab, 950 N. Cherry Avenue, Tucson, AZ 85719-4933, USA

Tim Sacco
NSFs NOIRLab, 950 N. Cherry Avenue, Tucson, AZ 85719-4933, USA

LSST-Discovery Alliance Catalyst Fellow, The School of Sociology at the University of Arizona

Dorian Russell
American Rescue Plan Act (ARPA), Washington County (United States), USA

IOP Publishing, Bristol, UK

ISBN   978-0-7503-4906-2 (ebook)
ISBN   978-0-7503-4904-8 (print)
ISBN   978-0-7503-4907-9 (myPrint)
ISBN   978-0-7503-4905-5 (mobi)

DOI   10.1088/2514-3433/ad2174

Version: 20240501

AAS–IOP Astronomy
ISSN 2514-3433 (online)
ISSN 2515-141X (print)

British Library Cataloguing-in-Publication Data: A catalogue record for this book is available from the British Library.

Published by IOP Publishing, wholly owned by The Institute of Physics, London

IOP Publishing, No.2 The Distillery, Glassfields, Avon Street, Bristol, BS2 0GR, UK

US Office: IOP Publishing, Inc., 190 North Independence Mall West, Suite 601, Philadelphia, PA 19106, USA

Dara Norman:

This ebook is dedicated to my amazing family, especially Knut, Tyra and Emil, and to the ancestors, on whose shoulders I stand.

Tim Sacco:

Dedicated to my partner Mahala and my son Sasha, both of whom fill our home with hope and love. Thanks for keeping me grounded in this wild world.

Dorian Russell:

This ebook is dedicated to the queer elders and chosen family who have shaped my potential, as well as my dearest Mimi and Drježdźany.

Contents

Preface

Practitioners of astronomy and astrophysics are dedicated to growing our knowledge of the Universe and all its contents by understanding the physics of waves, whether electromagnetic or gravitational. Over centuries, humans have been fascinated with the stars, the planets, the Moon, etc, and wanted to know more about what they are and how they have come to be. The desire to gain this knowledge about the Universe and our place in it is not limited to any one group of people in any one place or at any single time—it is a universal desire. However, the privilege to pursue an interest in astronomy and astrophysics as a professional career is not easy. It takes a keen interest, a strong desire and even some aptitude to pursue the discipline as a professional (or even as an amateur!). Any of those who are willing and want to invest their time in pursuit of making a career in astronomy should have the opportunity to do so regardless of their gender, race, sexuality, or other personal identity traits. However, it is well recognized that disparities in pursuing science, technology, engineering, and math (STEM), and more specifically astronomy and astrophysics, disciplines as a career exist. If we make the effort to look, we can see these disparities at our conferences, in our departments, and in our collaborations, just as we can see these disparities in our classrooms.

This book is about making time for the effort to look at the ways in which the current culture of science, of which astronomy and astrophysics are a subset, with its limited view of what is valued and narrowly dogmatic norms and traditions, have conspired to keep out many of those with the interest, desire, and aptitude to pursue the career. The goals of this book are to help readers see the ways in which we can dismantle a culture of exclusion and singularly competitive practices, and replace it with a culture that favors and rewards collaboration and inclusion; a culture that recognizes that aptitude is ubiquitous, if we take the time to identify it, but that opportunity is not; a culture that understands that as we identify capability in the places we have not looked for it before, we will actually unlock creativity and innovation that can lead to more knowledge, greater comprehension, and better science. This book serves to provide readers with an understanding of key disparities within the field, examples of successes and shortfalls in existing efforts, and how changes to the structure of the way science is performed can improve accountability in achieving equity in research opportunity and success.

Contributors to this book have called out the principles of equity, diversity, inclusion, accessibility, belonging, justice, and equality as they see fit to make their points about the culture of the field and how it needs to evolve. As editors, we have not modified how contributors express these sentiments. As a result, the reader will see DEI, EDI, DEIA, IDEA, EEID, DEIB or DEIJ mentioned. However, readers should note what the author is choosing to emphasize as they relay information about how to make the field more open, welcoming and fair.

The time for this book is now because the wider astronomical community has begun to recognize the advantages of being a field of science that is more welcoming, open and collaborative, as well as the justice in acknowledging and breaking down systemic barriers. Evidence of these changing attitudes is written about in this book and includes departmental DEI activities, written position (white) papers, the recommendations put forth by high level advisory committees, and the building of incentives into institutional structures that move the community in a direction of openness and inclusion.

The time for this book is also now because in just a few short years since the STEM community engaged in large scale, reflective activities like regular DEIA journal club discussions, conferences dedicated to more equitable research environments that would benefit the whole of the astronomical workforce[1] and #ShutDownSTEM,[2] the backlash to these efforts has arrived. This backlash is dedicated and political, but it is not new or unique. As usual, this backlash requires us to think in a false dichotomy that proposes that increased inclusion for some means decreased opportunities for others, instead of recognizing that increased inclusion will advantage all.

The audience for this ebook is intended to be everyone with an interest in STEM or astronomy and astrophysics; anyone with an interest in how we can grow the STEM workforce and make it more accessible and inclusive. Although astronomy examples are used as the backdrop, issues like systemic exclusion of some from science institutions, toxic culture in collaborations, and combating the growing anti-science sentiment in society are concerns that anyone with an interest in science can relate to.

The intended audience for this book may already know a lot about DEIA work in the areas of STEM or may know very little about these topics. This volume starts with highlighting the lived experiences of those in the field. We allow them, in their own words, to describe how they have found the past culture, recent strides made through the acknowledgement of bias and barriers, and their optimism for the future, if we are able to continue on a path of broad inclusion of all those interested in participating in astronomy careers. These stories may be entirely new to some of the intended audience, while others may recognize themselves in some stories but not others. The point, however, is to give these stories a voice before we lay out the stark statistics that reflect the concerns of who is involved in, and able to be involved in, professional astronomy.

A particular focus of this book is on the role of inter-departmental and inter-institutional collaborations for providing a catalyst for the evolution of professional astronomy culture. Astronomy and astrophysics are far from singular as STEM disciplines that rely on large collaborations to advance the broad goals of the field, i.e., to understand the Universe and its contents. However, given the requirements for collective facilities and infrastructure (i.e., large telescopes, satellites, computing power, etc) to pursue these lofty goals, resources must be pooled and so must be of benefit to the collective. Without some broad consensus, the community would be pulling in so many directions that it would not be possible to construct and deploy the needed facilities. In recent years, the size, complexity and importance of

---

[1] https://outerspace.stsci.edu/display/IA2
[2] https://science.mit.edu/shutdownstem/

astrophysical collaborations have been growing. It is through these collaborations that groups of astronomers cross pollinate, sharing expertise and resources across institutions. They also provide an opportunity for early-career researchers to become recognized and a vehicle for passing on the best, as well as the worst, values, norms and traditions of the culture. We focus on discussion about collaborations in this book because of their potential for driving systemic change throughout the field.

In this book we have attempted to highlight a range of voices, however we recognize that we have not been all-inclusive of every possible voice that there is to expose. Indeed, that would be impossible in the space of only 16 articles! Thus we recognize that this volume will not be all things to all people (it is not trying to be). However, we have contributors with intersectional identities, as well as those from outside of astronomy and astrophysics in order to capture a wide range of perspectives. We hope that we will have illuminated the many issues that exist in the field of astronomy and astrophysics with respect to workforce, highlighted some examples of how teams of astronomers have tried to tackle a few of these issues and concerns, and exposed ways in which the broader ecosystem of the field can incentivize and hold the astro community accountable for making cultural changes in support of science and the broadening of the scientific workforce. We hope that this book not only inspires the reader to identify ways that they can make changes within their own collaboration, department or institution to be more inclusive and equitable, but also that they identify ways to be more influential on these topics within the broader context of the field, STEM, and the promotion of science in society.

Lastly, we thank our contributors for sharing their stories, insights, expertise, and knowledge to help all of us understand our place in the Universe of ideas.

Editor biographies

Dara Norman

Dr. Dara Norman is the Deputy Director of the Community Science and Data Center at the NSF National Optical and Infrared Astronomy Research Laboratory (NOIRLab) in Tucson, AZ. Her research interests include the study of active galactic nuclei as a phase of galaxy evolution. She has served as AURA/NOAO Diversity Advocate, on the governing board of the American Astronomical Society, where she chaired the task force that revised the society's Ethics Code. She has been an active member of the AAS's Committee on the Status of Minorities in Astronomy (CSMA); chair of the astronomy and astrophysics section of the National Society of Black Physicists; and co-organizer of the 2015 Inclusive Astronomy Conference. She is a mother of two and spouse of one.

Tim Sacco

Dr. Timothy Sacco is an LSST-Discovery Alliance Catalyst Fellow in the School of Sociology at the University of Arizona. His research interests include the study of work and organizational dynamics in science. From 2021 until 2023, he was a postdoctoral fellow at NSF National Optical and Infrared Research Laboratory (NOIRLab) in Tucson, AZ, where he worked on the U.S. Extremely Large Telescope Program's Research Inclusion Initiative. While at NOIRLab, he worked alongside Dr. Norman to develop a Toolkit of Collaborative Practice, a resource designed to help Astronomers construct equitable and inclusive research collaborations. He has been an active member of the American Sociological Association's section on Science, Knowledge, and Technology. In 2017, he and his coauthors presented their research on inclusive collaborations to the White House Office of Science and Technology Policy. He currently lives in central New York with his partner and son.

Dorian Russell

Dorian Russell (they/them) is a public policy professional with advocacy and work experience in local, state, and federal government. Trained as an environmental scientist, Dorian's interdisciplinary career has focused on interpreting and applying science for decision makers and community. At the time of this book's writing, they serve as program administrator for Washington County, Oregon guiding the use of $117M across 50+ pandemic response and recovery programs in public health, workforce development, broadband access, and more. Prior to the COVID-19 pandemic, Dorian's research and advocacy activities focused on addressing workforce barriers for scientists and engineers, notably those who identify as BIPOC,

LGBTQIA2S+, and people with disabilities. In their free time, Dorian serves on the City of Hillsboro's Library Board advocating against book bans and for broadening resource accessibility. Formerly a homeless youth, they are also proud to serve on the Executive Board for HomePlate Youth Services, the only nonprofit providing both street outreach and drop-in services for homeless youth in their community. Dorian has been recognized nationally by Engaging Local Government Leaders (ELGL), the White House Council on Women and Girls, and others for their leadership in equity, diversity, and inclusion (EDI) strategies across sectors.

List of Contributors

Rachael L. Beaton
Space Telescope Science Institute, Baltimore, MD, 21218, USA
Department of Astrophysical Sciences, Princeton University, Princeton, NJ, USA
The Observatories of the Carnegie Institution for Science, San Diego, CA, USA

L. Michelle Bennett
LMBennett Consulting, LLC, Potomac, MD, USA

Federica Bianco
Department of Physics and Astronomy, University of Delaware, Newark, DE, USA
Jr. School of Public Policy and Administration, University of Delaware Joseph R. Biden, Newark, DE, USA
Data Science Institute, University of Delaware, Newark, USA
Center for Urban Science and Progress, New York University, Brooklyn, NY, USA

Heather Bloemhard
Office of Federal Relations, Vanderbilt University, Nashville, TN.

Robert D. Blum
Rubin Observatory NSF NOIRLab, Tucson, AZ, USA

Rosaria (Sara) Bonito
INAF—Osservatorio Astronomico di Palermo, Palermo, Italy

The Committee on INclusiveness in SDSS (COINS)

Megan Donahue
Michigan State University, East Lansing, MI, USA

Jessica Esquivel
Fermilab, Batavia, IL, USA

Kathryn (Kat) G. Gardner-Vandy
Oklahoma State University, Stillwater, OK, USA

Ranpal Gill
Rubin Observatory NSF NOIRLab, Tucson, AZ, USA

Jarita Holbrook
The University of Edinburgh, Edinburgh, Scotland

Rachel Ivie
American Institute of Physics, Melville, NY, USA

Amy M. Jones
Space Telescope Science Institute, Baltimore, MD, USA

A. J. Link
AstroAccess, Washington, DC

Andrés A. Plazas Malagón
Kavli Institute for Particle Astrophysics and Cosmology, Stanford, CA, USA
SLAC National Accelerator Laboratory, Menlo Park, CA, USA
Department of Astrophysical Sciences, Princeton University, Princeton, NJ, USA

Lindsey Malcom-Piqueux
California Institute of Technology, Pasadena, CA, USA

Wil O'Mullane
Rubin Observatory NSF NOIRLab, Tucson, AZ, USA

Dara Norman
NSF NOIRLab, Tucson, AZ, USA

Dorian Russell
Program Administrator County Administrative Office, Washington County, OR, and City of Hillsboro Library Board, Hillsboro, OR, USA

Tim Sacco
NSF NOIRLab, Tucson, AZ, USA

LSST-Discovery Alliance Catalyst Fellow, The School of Sociology at the University of Arizona.

Daniella M. Scalice
NASA Ames Research Center, Moffett Field, CA, USA

Alysha Shugart
Rubin Observatory NSF NOIRLab, Tucson, AZ, USA

Rachel Street
Las Cumbres Observatory, Goleta, CA, USA

Aprajita Verma
Sub-department of Astrophysics, University of Oxford, Oxford, UK

Susan White
American Institute of Physics, Melville, NY, USA

Part I

Centering the Stories

Introduction to Part I: Centering the Stories

The goal of Chapter Part I is to provide the reader with an understanding of the historical and systemic nature of bias in STEM fields and the need to embrace the principles of diversity, equity, inclusion, and access. The articles in this part communicate this need through the experiences of those colleagues that have first-hand knowledge of and familiarity with marginalization in the field of astronomy and astrophysics, or other closely related STEM fields. Several of these articles explore the marginalization of STEM professionals in an intersectional way, that is, along more than one axis of their identity, e.g., Black and female. Their accounts scrutinize the ways the norms, traditions and values of the science culture have perpetually, over several decades, excluded or marginalized some groups of people from fully participating in the broad astronomy enterprise. This enterprise includes, not only scientific research, but also other adjacent technical careers that contribute to space discoveries.

These testimonies are representative of the concerns of many marginalized groups with the current state of the profession. Each article identifies strategic paths forward to mitigate past harms. For some, embracing that sense of belonging has spurred them to activism within the field and for others it has meant leaving the field altogether and working to change systems of oppression from another vantage.

In *"Detecting the Signal Amidst the Noise..."*, the author explores the long history of systemic racial discrimination in science and the toll that it takes on those who continue to persist in the field despite a myriad of barriers. *"It All Starts with Relationships..."* explores the current state of astronomy culture and the need for a more humble and collaborative approach to existing with local indigenous communities. *"An Accessible Future"* challenges the stereotypical notions of disability with a reorientation of how accessibility should look. After centering the stories and perspectives of the authors above, we end the chapter with statistics that demonstrate just who is being left behind in the article, *"Are We Missing Out?"*

This part sets the stage for the reader to become more fully aware of the complexities of the barriers to a sense of belonging, inclusion, equity and accessibility that exist within astronomy and astrophysics, especially for those who are at the intersection of multiple identities.

An Astronomical Inclusion Revolution
Advancing Diversity, Equity, and Inclusion in Professional Astronomy and Astrophysics
Dara Norman, Tim Sacco and Dorian Russell

Chapter 1

Detecting the Signal Amidst the Noise: Ambient Exclusion as a Barrier to Advancing Diversity in Physics and Astronomy

Lindsey Malcom-Piqueux

Representational inequities based on race, gender, and the intersection of these two, continue to persist in physics and astronomy careers, despite decades of efforts—policy, disciplinary, institutional, and philanthropic—to abate them. Though incremental progress has been made for some historically excluded, marginalized, and minoritized populations, as evidenced by the growing proportional representation among physics and astronomy bachelor's degrees earners over the past few decades, concomitant gains in graduate degree attainment and workforce participation have not been realized.

The barriers to realizing equity in physics and astronomy are numerous, complex, and operate at the intrapersonal, interpersonal, institutional, and systemic levels (National Academies of Science, Engineering, and Medicine [NASEM] 2021, 2022). Though research has identified multiple factors that support or hinder success in physics and astronomy education and careers, access to such supports and experiences of these barriers vary across dimensions of identity and demographic groups. Consider the following example: a strong STEM identity is now understood to be critically important to success in these fields broadly (Carlone & Johnson 2007; Seymour & Hunter 2019), and in physics and astronomy specifically (American Institute of Physics 2020). Recognition, encouragement, and validation from faculty, peers, and others in the discipline are vital to the formation of a strong STEM identity, particularly for members from historically excluded and marginalized populations. However, the gender-based stereotypes that a white woman in physics must navigate and overcome as her STEM identity is shaped are distinct from the intersectional racialized and gendered stereotypes that a Black woman in physics faces as her STEM identity is forming, given how racism, anti-blackness, and misogyny operate within the broader social context, higher education, and the

doi:10.1088/2514-3433/ad2174ch1

scientific disciplines. The overlapping, yet distinct, histories of identity-based exclusion in STEM (including physics and astronomy) necessitate that approaches to mitigating the barriers to advancing diversity and equity in these fields are crafted in identity-conscious ways (Dowd & Bensimon 2015).

Given that the "differentiated histories" of exclusion (Dowd & Bensimon 2015, p. 59) require nuanced, identity-conscious approaches to *inclusion*, this piece focuses on African Americans in physics and astronomy. Drawing upon theory and research, I explore how signals of exclusion present within learning, research, and working environments in physics and astronomy act as barriers to equity for African Americans and, by extension, undermine the broader goal of diversifying the physics and astronomy workforce. I identify multiple forms of ambient exclusion—that is, feelings of discomfort due to signals within one's environment that they are not accepted, not valued, and that they do not belong—that African Americans encounter within physics and astronomy. Addressing these signals of exclusion will help to create the conditions for equity-minded[1] practice within physics and astronomy.

As a Black woman who, after being educated at historically white, STEM-focused institutions, left the field of planetary science, my positionality shapes my treatment of this topic. Though I draw on some aspects of my lived experiences throughout this piece, the purpose of this approach is not to generalize based on what I have faced; rather, I invoke personal stories to individualize what research has long told us about how the absence of belonging acts as a barrier to African American student success in STEM and to the entry and retention of African Americans in STEM-focused careers. I begin by briefly highlighting the inequities experienced by African Americans in physics and astronomy.

1.1 The State of Equity for African Americans in Physics and Astronomy

Multiple identity groups remain underrepresented within physics and astronomy; however, the equity challenges experienced by African Americans are particularly troublesome (AIP 2020). While African Americans' share of all earned bachelor's degrees in astronomy grew from 0.5% in 2000 to 2.2% in 2020, this represents a numerical increase of just sixteen additional degree earners (going from two astronomy bachelor's degrees in 2000 to 18 in 2020) (National Center for Education Statistics [NCES] 2022). The 'small *n*'s' in astronomy are attributable in part to the relatively limited number of bachelor's degree programs in this field in the USA, and their notable absence from HBCUs. These large-scale, structural factors make it difficult to assess the state of equity in astronomy examining degree attainment data alone. In physics, however, trend data on

---

[1] Being *equity-minded* entails noticing racial inequities in outcomes, understanding these inequities in the sociohistorical context of systemic racism and exclusion within academia, and feeling a sense of personal and institutional responsibility to eliminate those inequities by critically reassessing taken-for-granted routines, norms, and practices (Dowd & Bensimon 2015; Center for Urban Education n.d.; McNair et al. 2020).

bachelor's degree attainment are more instructive. Though the number of bachelor's degrees in physics earned by African Americans increased from 152 in 2000 to 253 in 2020, African Americans' share of all earned physics bachelor's degrees decreased from 4.5% to 3.3% over that same period (NCES 2022). It is also important to note that the numerical increase was nearly entirely attributable to a greater number of HWIs graduating a single African American in physics (AIP 2020; NCES 2022). Is this really progress? From an equity perspective, the answer is an emphatic no.

African Americans' declining share of physics bachelor's degrees over the past two decades illustrates that though the output of undergraduate physics degree programs grew, this expansion did not benefit the Black community in the same manner as other demographic groups. Similar patterns hold for graduate degrees in physics and astronomy. Between 2000 and 2020, the total number of earned master's degrees in physics increased by 71% and the total number of earned doctorates in physics increased by about 50% (NCES 2022). Over this same period, the number of master's degrees in physics earned by African Americans increased by just 9% and the number of doctorates in physics earned by African Americans remained flat (NCES 2022). In short, while the number of physics degrees conferred by U.S. institutions grew significantly over the past twenty years, African Americans did not experience the same kind of expanded access to degrees in these fields. The data presented above are indicative of growing inequities in terms of access to physics and astronomy degrees, and by extension, careers in these fields.

As described in the extensive report of the American Institute of Physics National Task Force to Elevate African American Representation in Undergraduate Physics and Astronomy (TEAM-UP) (AIP 2020), the inequities in physics and astronomy experienced by African Americans are attributable to a constellation of factors, some of which are structural and others that operate at the institutional and departmental levels. TEAM-UP's evidence-based recommendations call for faculty, institutions, and professional societies to promote African American student success in physics and astronomy by fostering inclusive educational practice, climate, policies, and support structures.

Maintaining a strong physics identity and sense of belonging remain important over the course of physics and astronomy careers. However, the predominantly white and male professional environments in which African American physicists and astronomers work can undermine identity and belonging. For example, in their study of more than 500 STEM faculty at a research university, Blair-Loy & Cech (2022) found that racially minoritized faculty must navigate a gauntlet of stereotypes and negative assumptions about their competence, the "quality" of their research, and the reasons for which they were hired (i.e., being labeled as a diversity hire). Given paucity of African Americans in the physics and astronomy workforce, these exclusionary "push factors" within academic and research environments can exacerbate feelings of isolation and contribute to negative experiences along their career paths (e.g., Prescod-Weinstein 2021; NASEM 2022; Woods & Walker 2022). Intentional, collaborative efforts to promote inclusion and foster belonging by critically reassessing existing practices and reforming the climate and culture of academic, research, and professional environments offer a path forward.

1.2 Inclusive Practice and Belonging among African Americans in Physics and Astronomy

Inclusive practices are those that intentionally engage diversity in a manner that ensures all members of an organization possess a sense of belonging, feel validated by their colleagues, and are viewed as valued members of a team working toward a shared goal. The TEAM-UP report highlights the importance of inclusion and belonging to African American student success, and the task force's recommendations identify faculty, physics and astronomy departments, and professional societies as critical to promoting students' sense of belonging. African American professionals in the physics and astronomy workforce, particularly those in the early-career stage, have also highlighted the importance of inclusive environments to their retention and success in their own careers (Isler 2015; Alexander 2021; Prescod-Weinstein 2021; Blair-Loy & Cech 2022; NASEM 2022; Woods & Walker 2022). Inclusive environments are those in which African American students' and professionals' identities as knowers and doers of physics and astronomy are validated by their peers, faculty, colleagues, and others. Where diversity is just tolerated, goes unacknowledged, or is seen as counterproductive to the goals of an organization, inclusion cannot exist. Without inclusion, diversity certainly won't increase, and what little exists is often not sustainable.

The TEAM-UP report calls for physics and astronomy faculty, departments, and institutions to build their capacity for inclusive practice, create counterspaces, establish norms of respect and inclusion, and conduct regular assessments of departmental climate to promote a sense of belonging among African American students (AIP 2020). Within physics and astronomy working environments where team science is prevalent, inclusive leadership and mentorship strategies emphasizing the importance of mitigating the effects of conscious and unconscious biases, establishing psychological trust, and interrupting microaggressions and other exclusionary interactions help to create the conditions for inclusion. In fact, NASA and other federal agencies are beginning to mandate that grantees create and implement inclusion plans as a condition for funding. While these recommendations offer actionable strategies to advance inclusion in physics and astronomy, the physics and astronomy community should also critically reflect upon how taken-for-granted practices, norms, and beliefs within the discipline can send signals to African Americans that they do not belong in these fields, degree programs, and careers. As part of the disciplinary cultural milieu, these ambient signals of exclusion are embedded within academic, research, and professional environments in physics and astronomy, often going unquestioned and unexamined by dominant groups within these fields. Because belonging can only be fostered when ambient exclusion is recognized and dismantled, identifying these harmful signals is essential. Once those signals are identified, equity-minded collaboration within the physics and astronomy community can be employed as an approach to enact inclusion with intention.

In the following section, I describe common forms of ambient exclusion present within physics and astronomy learning, research, and professional environments as they are currently constructed. By examining the disciplinary conditions that contribute to racial inequities, the physics and astronomy community will better understand what transformative steps can be taken to advance equity for African Americans in these fields.

1.3 Signals of Ambient Exclusion as Equity Threats

1.3.1 Framing Inequities in Physics and Astronomy Experienced by African Americans Solely as a Matter of Underrepresentation

African Americans (and other racially minoritized populations) are often described as "historically underrepresented" within physics, astronomy, and other STEM fields. While this is an accurate mathematical description—indeed, African Americans comprise a smaller share of physics and astronomy degree holders and the physics and astronomy workforce than their share of all degree holders, the STEM workforce, and the general population—characterizing the issue solely as a matter of "underrepresentation" belies the insidious, historical exclusion that created the conditions for the present demographic reality. This goes beyond semantics; language is important as it reflects and shapes how we make sense of the issue, and by extension, our understanding of what should be done and who has the responsibility to act (Dowd & Bensimon 2015; McNair et al. 2020).

When confronting any challenge, individuals employ a cognitive frame or mental schema that helps them make sense of the problem. The cognitive frame shapes an individual's understanding of the problem, including its causes, effects, and the actions needed to address the problem (Gioia & Poole 1984). Framing the persistent inequitable access to physics and astronomy degrees and careers experienced by African Americans as an issue of underrepresentation alone provides no clear directive on what the causes are, what actions should be taken, the appropriate points of intervention, and who bears the responsibility of acting to redress the issue. At best, this characterization frames inequities as an inevitable outcome, akin to an unfortunate, yet uncontrollable natural disaster (Dowd & Bensimon 2015; McNair et al. 2020). At worse, framing inequities solely as "underrepresentation" leaves room for deficit-minded[2] actors to claim that inequities stem from what African Americans are perceived to "lack" (e.g., lacking interest in physics and astronomy, lacking academic preparation or requisite knowledge to succeed in these fields, lacking awareness of physics and astronomy career paths). Blaming equity challenges on the very individuals and communities experiencing them insinuates that they deserve what they get—a cruel irony given the long history of racialization and stratification in the U.S. The educational system and scientific enterprise were constructed to provide inequitable opportunity for exposure to, preparation for, and encouragement within all STEM fields, including physics and astronomy (Malcom-Piqueux 2021). Characterizing the inequities experienced by African Americans in physics and astronomy as an issue of "underrepresentation" alone ignores that these

---

[2] In this context, *deficit-mindedness* involves viewing racial inequities as originating from perceived deficits or characteristics of the very individuals, families, and communities experiencing those inequities (Bensimon 2007).

inequities were created by design through systemic and institutional racism, as well as discriminatory policies and practices that purposefully excluded African Americans and other minoritized groups from educational and professional opportunities in these and other STEM fields (Malcom-Piqueux 2021).

1.3.2 Maintaining Racialized Expectations of Who Will Show Up in Physics and Astronomy Academic and Professional Environments

Given that historical exclusion of African Americans and other minoritized groups from physics and astronomy has led these fields to be demographically homogenous, it is not surprising that, as in other physical sciences, the stereotypical physicist and astronomer is white and male (Nosek et al. 2009; Cheryan et al. 2015). Though such stereotypes are relatively safe bets given the demographic makeup of those holding doctorates in physics and astronomy, these widely shared cultural perceptions often hinder the diversification of these fields and other STEM disciplines (Cheryan et al. 2015). Such stereotypes have the potential to be even more harmful when perpetuated by those inside the field because they limit expectations of who "will show up" in physics and astronomy academic, research, and professional environments. When someone outside of those expectations arrives, the dissonance between what is observed and prior expectations sends clear signals about who and who does not belong.

To illustrate this point, consider the following examples.

In 1973, on the first day of the semester at Penn State, a Black graduate student walked into his solid-state physics class. With an enrollment of around ten students, the course was quite small. As he approached one of the desks, the professor looked at him with questioning eyes. "Are you sure you're in the right place?", the professor asked. The student responded affirmatively, but the professor continued to look at him quizzically. "What major are you?" the professor asked the Black graduate student. "Computer science," the student replied. "Well...I guess I can see the connection" the professor responded. The student, keenly aware that he was the only person of color in the class, sat down for lecture.

On a Saturday morning in 2000, a Black woman was walking down the familiar halls of Building 4, to get to the classroom to sit for the Physics GRE subject test. The senior, along with many other of her classmates at MIT, was preparing to apply to graduate school—the next logical step to pursue her desired career in Planetary Science. She had spent the entire summer studying, but it wasn't enough to keep her nerves at bay. The student planned to arrive at the room an hour before the start time to secure her preferred seat and to settle in before the exam. As the Black woman entered the room, she noticed a few students were already there. She smiled and tried to make eye contact, when one of them said to her, "Umm...this is the room for the Physics GRE." It took the Black woman a split second to realize that this person was directing his comment towards her. Flustered, she responded, "I know." She headed to the first seat available and sat down, her heart beating fast and heat spreading from her face to her ears. She asked herself why that student would direct that comment to her, in that tone and with the way that he emphasized the word, 'Physics'. He didn't say it to any of the other students who came into the room after she did. She wondered what it was about her *that made it look like she was lost or in the wrong place. She knew the answer. As the room filled, it became glaringly obvious. She was one of a few women and the only Black person in the room of dozens of students. According to this white, twentysomething-year old man who had appointed himself gatekeeper, she didn't belong.*

Though these incidents occurred 27 years apart, they both reflect the ongoing struggle of African Americans in physics, astronomy, and related disciplines to find belonging in an environment where the prevailing assumption has long been that we do not. As Sara Ahmed (2012) explains, "that the arrival of some bodies is more noticeable than others reveals an expectation of who will show up" (p. 42). In community conversations, journal articles, memoirs, and even TEDtalks, African Americans in physics and astronomy at all career stages have recounted similar instances during which their presence so defied the expectations of those around them that they were met with stares of disbelief, interrogated about why they were there, or told that they were in the wrong place. The examples above illustrate what occurred when my father and I defied others' racialized expectations of who belonged in physics, though to very different ends. After completing his graduate degrees in computer science and physics, my father had a decades-long career in these fields. By contrast, I left my doctoral program in planetary science full of self-doubt fed by the inescapable signals that I did not belong.

Given the importance of belonging to success for African Americans in physics and astronomy, being forced to legitimize your presence due to your defiance of others' racialized expectations is a signal of ambient exclusion and an equity threat. It is difficult to feel that you belong if you are constantly being told that you do not. Maintaining racialized expectations of "who will show up" in physics and astronomy academic and professional environments is problematic because such expectations are rooted in whiteness and undermine the sense of belonging of those who fall outside of those expectations. Indeed, my own discomfort with my "failure to fit" over the course of my undergraduate and graduate years led me to leave the field of planetary science entirely (Ahmed 2012, p. 41).

1.3.3 Ignoring that Physics and Astronomy are Human Endeavors in Which Bias, Racism, and Discrimination Shape the Experiences and Outcomes of African Americans and Other Minoritized Groups

Physics and astronomy, like many other STEM fields, tout objectivism as a foundational principle of its disciplinary ideology (e.g., West 1960; Pearson 1978;

Whitten 1996; Crotty 1998). Objectivity, or the pure state of the mind untainted by emotion, bias, pre-held beliefs, and opinions, is believed to be both attainable and desirable according to the dominant ideology of science. Reverence for and the purported adherence to objectivity are deeply entrenched within the disciplinary cultures that operate within physics and astronomy academic and professional environments (Traweek 1988). As students are socialized into these disciplines, objectivity is elevated, and subjectivity is deemed to have no place within the realm of physics and astronomy. A logical extension of this perspective is that the process of doing physics and astronomy, and the physicists and astronomers engaged in this work are also objective, insulated from the social forces that operate in the context in which science is done. Described by Traweek (1988) as the "culture of no culture", this problematic assertion fails to recognize science as a human endeavor and belies a growing body of research that illustrates how bias, racism, and discrimination shape the outcomes and experiences of African Americans and other racially minoritized groups in these fields (p. 162).

Feelings of isolation (Tate & Linn 2005), hypervisibility, and invisibility (Kohli 2015) are commonly reported by African Americans in physics and related fields. Several researchers have examined the racialized experiences of African Americans in physics and other STEM disciplines, noting the ways in which Black people in these fields must navigate the complexities of racial prejudice, stereotypes, and imposter phenomenon, while resolving perceived conflicts between their intersecting social identities and their identity as physicists (e.g., Fries-Britt et al. 2010; Fries-Britt & Holmes 2012; Fries-Britt et al. 2013; Rosa & Mensah 2016; McGee 2020). Additionally, African American students in physics and other STEM fields face racial microaggressions, overt racism, and discrimination from peers and faculty more often than other demographic groups (e.g., Johnson 2007; McGee & Martin 2011; Dortch & Patel 2017; McGee 2020; Park et al. 2022; Rodriguez et al. 2022).

Bias, racism, and discrimination also act as barriers for African Americans within the physics and astronomy workforce (Porter 2019; Eaton et al. 2020; Porter et al. 2020; Prescod-Weinstein 2020). Eaton et al. (2020)'s experimental study found that physics faculty members' assessments of prospective postdoctoral candidates were biased on the basis of race and gender. Faculty were asked to evaluate CVs of potential postdoc candidates for their hireability and competence. The CVs were *identical* except for the gendered and racialized names of the fictional candidates. Physics faculty rated candidates with male identified names higher on hireability and competence than candidates with female-identified names. Physics faculty also rated Asian and White candidates as more hirable and more competent than Black and Latinx candidates. Though Eaton et al.'s (2020) finding of racial and gender bias in hiring decisions mirrors previous studies of hiring within the general workforce (e.g., Bertrand & Mullainathan 2004), a notable difference is that the postdoctoral candidates being evaluated had already earned a doctorate. Thus, despite meeting an "objective" benchmark of completing the PhD, the Black postdoctoral candidates were subject to racial bias and discrimination. African American professionals in physics and astronomy often face the minority-meritocracy paradox (Blair-Loy & Cech 2022) where Black scientists must go above and beyond to prove their scientific mettle to overcome a racially biased hiring process only to have colleagues to question their competence and view their science as subpar once they are hired. Then, after succeeding despite these barriers, Black scientists often report that their scientific accomplishments are discounted and attributed to unfair advantages that they received due to their race (e.g., being a diversity hire). This farcical narrative belies objectivity and logic—two values that are elevated in science. Nonetheless, the minority-meritocracy paradox is commonly experienced by racially minoritized scientists and engineers (Blair-Loy & Cech 2022).

Clinging to objectivity as canon involves ignoring how oppressive forces like racism and anti-blackness inform prevailing assessments of who possesses the competence, intelligence, and drive to succeed in physics and astronomy. In addition to denying the lived experiences of African Americans in physics and astronomy, dogmatic adherence to objectivity as an organizing principle of physics and astronomy requires that we dismiss the preponderance of social science evidence about how bias, racism, and discrimination operate within these fields. We cannot solve what we will not face.

1.3.4 Practicing Uncritical Hero Worship of Physicists and Astronomers

Hero worship—the elevation and adoration of an individual who is viewed to have made exceptional contributions to the advancement of society—is present within STEM fields, but particularly in physics, astronomy, and mathematics (Hottinger 2016; Vazire 2017; Posselt 2020). While hero worship rose to prevalence during the Victorian age (e.g., Carlyle as cited in Houghton 1957/2014), the practice of recounting "the history of the world" through "the biography of great men" is still common. In physics and other STEM fields, hero worship often takes the form of elevating individual scientists who have made groundbreaking discoveries that are foundational to those disciplines to genius status. Learning the names of these scientific heroes and acquiring the ability to describe their contributions is part of the socialization process into the disciplinary culture. While there are certainly modern-day heroes in physics and astronomy, the idolatry of the greats of the past (e.g., Newton, Kepler, Millikan, Einstein, Feynman) continues, as evidenced by examining most introductory physics or astronomy textbooks.

Hero worship is a problematic practice for many reasons, some of which are related to scientific progress. For example, when an active scholar reaches elevated "hero" status, the accompanying fawning adoration of that researcher and their prior success can lead to their future work going unquestioned, skirting by the typical scrutiny of evaluative processes and the disciplinary community (Vazire 2017). Hero worship, particularly when it is uncritical, can also act as a form of ambient exclusion, threatening equity within scientific disciplines. The uncritical worship of figures from the past paint a homogenous portrait of the discipline. Given that the foundations of physics and astronomy were constructed in a social context within which racism and other forms of oppression were used to justify denying minoritized groups educational resources and opportunity, it is not surprising that the vast majority of the "heroes" being worshiped are white men. The contributions of others, while present, remain hidden from the dominant narratives of the emergence and advancement of physics and astronomy. Thus, white men are "invited to see themselves within these histories" and women of all races and racially minoritized people of all genders are "excluded from these histories" (Hottinger 2016, p. 12). Further, these "heroic" white men are often mythologized to have been innately brilliant, possessing raw talent rarely observed in the field. Thus, uncritical hero worship of figures perpetuates the lore that innate intellectual talent is critical to success within these disciplines. Due to our society's deeply entrenched racist, anti-Black, and gendered stereotypes, innate brilliance is less likely to be associated with African Americans than with white men (Leslie et al. 2015).

Uncritical hero worship within physics and astronomy can also send signals of ambient exclusion when the white, male supremacist beliefs and related actions of heroes are ignored or characterized as irrelevant to their legacies. Recent debates about the (re)naming of buildings, prizes, professorships, and fellowships on campuses across the nation illustrate how this occurs. As Sara Ahmed (2012) explains, "acts of naming, of giving buildings names, can keep a certain history alive: in the surroundings you are surrounded by who was there before" (p. 38). As research universities grew and NASA expanded throughout the 20th century, the visible history of physics, astronomy and related fields came into stark relief. Robert Millikan, John LeConte, Robert Lee Moore, Richard Feynman, and James Webb, among others, have been memorialized with buildings, prizes, and even instrumentation intended to further human knowledge. Though seen as worthy of honor for their contributions to physics and astronomy in the past, until recently, their contributions and influence on the *social* world went largely unnoticed. Due in large part to increased awareness of the racist, sexist, antisemitic, and anti-LGBTQ+ perspectives and positions of these "heroes" brought about by the activism of students, faculty, and professionals in the scientific community, many colleges and universities considered whether those names should remain in a place of honor. While each institution followed its own process to determine whether to remove or retain the names, debates about whether the problematic views of these individuals outweighed their contributions to science were rampant. Advocates of retaining the names of physicists, astronomers, and other scientists now widely known to have associated with white supremacy, misogyny, and homophobia hoped that the scientific and campus communities would be able to compartmentalize, walling off their scientific achievements from their deplorable worldviews. Requesting that minoritized groups overlook the de-humanization of and discrimination against people who shared their identities sends a clear message about what does and does not matter to a significant segment of the scientific community. Who would blame an African American physics student for interpreting the vocal opposition of some peers, faculty, and alumni from their department as a signal that they do not belong?

Exclusionary practices in physics and astronomy undermine efforts to achieve racial equity in these fields. As I argued above, many taken-for-granted disciplinary cultural norms, features within the institutional and departmental climate, and even the discourse about increasing diversity in physics and astronomy can send signals to African Americans and other historically excluded groups that they do not belong. To advance equity in physics and astronomy, these signals of ambient exclusion should be examined and addressed through intentional efforts. Transforming the learning, research, and professional environments to advance inclusion in physics and astronomy will require collaboration within these disciplinary communities across academia, industry, funding agencies, and professional societies.

1.4 Advancing Inclusion through Equity-Minded Collaboration

Given that the forms of ambient exclusion described above stem from entrenched disciplinary norms, cultural change is an essential component of the work that must be done to advance inclusion in physics and astronomy. Culture, by definition, is a collective project, co-constructed over time and reified by individual, institutional, and community practices, policies, and processes. Because culture is constructed and enacted collaboratively, changing culture must also be a group endeavor. Equity-minded collaboration provides one means to change the culture of physics and astronomy by rooting out its ambient and overt exclusionary elements to normalize diversity and inclusion.

Equity-minded collaboration involves stakeholders working together to remediate practices, policies, and processes that contribute to racial inequities. In this context, equity-minded collaboration to advance inclusion in physics and astronomy would engage professionals, researchers, educators, students and others in the community as well as professional societies, scientific and research organizations, funding agencies, and higher education institutions in efforts to identify and reform practices that contribute to racialized inequities in participation, representation, and outcomes in these disciplines. The engagement of all stakeholders is required to ensure that equity in physics and astronomy is prioritized, supported with resources, advanced through inclusive policies and practices, and encouraged with appropriate incentive structures. Equity-minded collaboration intended to further inclusion in astronomy and physics might explore the following questions:

- How can the community reframe diversity, equity, and inclusion work in physics and astronomy as essential to the future of these fields instead of being derided as "impure" or a distraction?
- How can professional societies, scientific and research organizations, and the academy ensure that physics and astronomy community members who contribute to inclusion goals are recognized, valued, and rewarded tangibly?
- How can governmental funding agencies and philanthropic organizations that support physics and astronomy research incentivize using inclusion plans as a space for collaboration between physics and astronomy researchers and DEI experts?
- How might reconceptualizing "merit" and "fit" in ways that prioritize experience in supporting diversity and equity and knowledge of inclusive practice lead to a more diverse physics and astronomy faculty? (e.g., Liera & Ching 2020).

Engaging in collaborative work to address these questions would spark generative conversations about the changes to policy and practice needed to create a more equity-minded culture in physics and astronomy. Examples of collaborative approaches intended to advance diversity and inclusion in astronomy are included in the Astro2020 report (NASEM 2021). It is important to ensure that such efforts center equity by attending to the ambient, covert, and overt forms of exclusion that currently exist within physics and astronomy. Fortunately, there are national-level initiatives that provide models of how to engage in critical self-assessment for the purposes of cultural change. For example, the American Association for the Advancement of Science (AAAS) SEA Change initiative engages institutions and departments in a collaborative effort to assess their current state of equity in STEM and to understand how policies, practices, and processes related to recruitment, hiring, retention, admissions, teaching, and mentoring have yielded those outcomes. The primary goal of this work is to identify points of intervention for more inclusive practice (AAAS 2022). The American Astronomical Society, American Physical Society and other professional societies and scientific organizations have also engaged in efforts to promote and proliferate inclusive approaches within physics and astronomy. Additionally, NASA, the National Science Foundation, and the Department of Energy are beginning to implement requirements related to advancing diversity and inclusion in their proposal processes. Clearly, critical actors within the physics and astronomy community see the need to make these disciplines more inclusive. To fully realize their aims, these collaborative efforts should also address all forms of exclusion—both the visible and hidden.

References

Ahmed, S. 2012, On Being Included: Racism and Diversity in Institutional Life (Durham, NC: Duke Univ. Press)

Alexander, S. 2021, Fear of a Black Universe: An Outsider's Guide to the Future of Physics (New York: Basic Books)

American Association for the Advancement of Science 2022, AAAS SEA Change Guiding Principles, https://seachange.aaas.org/about/principles

American Institute of Physics 2020, The Time Is Now: Systemic Changes to Increase African Americans with Bachelor's Degrees in Physics and Astronomy (College Park, MD: American Institute of Physics)

Bensimon, E. M. 2007, Rev. High. Ed., 30, 441

Bertrand, M., & Mullainathan, S. 2004, Am. Econ. Rev., 94, 991

Blair-Loy, M., & Cech, E. A. 2022, Misconceiving Merit: Paradoxes of Excellence and Devotion in Academic Science and Engineering (Chicago, IL: Univ. of Chicago Press)

Carlone, H. B., & Johnson, A. 2007, JRScT, 44, 1187

Center for Urban Education n.d., Equity Mindedness (Los Angeles, CA: Rossier School of Education, Univ. of Southern California) https://cue.usc.edu/equity/equity-mindedness/

Cheryan, S., Master, A., & Meltzoff, A. N. 2015, Front. Psychol., 6, 49

Crotty, M. 1998, The Foundations of Social Research (Thousand Oaks, CA: Sage)

Dortch, D., & Patel, C. 2017, NASPA J. Women High Educ., 10, 202

Dowd, A. C., & Bensimon, E. M. 2015, Engaging the "Race Question": Accountability and Equity in U.S. Higher Education (New York: Teachers College Press)

Eaton, A. A., Saunders, J. F., Jacobson, R. K., & West, K. 2020, Sex Roles, 82, 127

Fries-Britt, S., & Holmes, K. M. 2012, Black Female Undergraduates on Campus: Successes and Challenges, ed. C. Renée Chambers, & R. Vonshay Sharpe (Bingley: Emerald Group Publishing Limited) 199

Fries-Britt, S. L., Younger, T. K., & Hall, W. 2010, Managing Diversity: (Re)Visioning Equity on College Campuses, ed. E. T. Dancy (Lausanne: Peter Lang) 181

Fries-Britt, S. L., Johnson, J., & Burt, B. 2013, T.L. Strayhorn Living at the Intersections: Social Identities and Black Collegians, 21 (Charlotte, NC: Information Age Publishing)

Gioia, D. A., & Poole, P. P. 1984, Scripts in Organizational Behavior. AMR, 9, 449

Hottinger, S. N. 2016, Inventing the Mathematician: Gender, Race, and Our Cultural Understanding of Mathematics (New York: State Univ. of New York Press)

Houghton, W. E. 2014, The Victorian Frame of Mind, 1830–1870 (New Haven, CT: Yale Univ. Press) (Original work published 1957)

Isler, J. C. 2015, The New York Times, https://www.nytimes.com/2015/12/17/opinion/the-benefits-of-black-physics-students.html

Johnson, A. C. 2007, SciEd, 91, 805

Kohli, S. 2015, In 39 years, US Physics doctorates went to 66 Black women—and 22,000 white men, https://qz.com/432756/in-39-years-us-physics-doctorates-went-to-66-black-women-and-22000-white-men

Leslie, S. J., Cimpian, A., Meyer, M., & Freeland, E. 2015, Sci., 347, 262

Liera, R., & Ching, C. 2020, ed. A. Kezar, & J. Posselt Higher Education Administration for Social Justice and Equity: Critical Perspectives for Leadership (Milton Park: Routledge)

Malcom-Piqueux, L. E. 2021, Transformation in the U.S. Higher Education System: A Critical Historical Overview and the Implications for Equity, National Academies of Sciences, Engineering, and Medicine, Board on Science Education https://www.nationalacademies.org/event/10-21-2020/imagining-the-future-of-undergraduate-stem-education-symposium

McGee, E. O. 2020, Educ. Res., 49, 633

McGee, E. O., & Martin, D. B. 2011, Am. Educ. Res. J., 48, 1347

McNair, T., Bensimon, E. M., & Malcom-Piqueux, L. E. 2020, From Equity Walk to Equity Talk: Expanding Practitioner Knowledge for Racial Justice in Higher Education (Hoboken, NJ: Jossey-Bass)

National Academies of Sciences, Engineering, and Medicine 2021, Pathways to Discovery in Astronomy and Astrophysics for the 2020s (Washington, DC: The National Academies Press)

National Academies of Sciences, Engineering, and Medicine 2022, Advancing Diversity, Equity, Inclusion, and Accessibility in the Leadership of Competed Space Missions (Washington, DC: The National Academies Press)

National Center for Education Statistics, Integrated Postsecondary Education Data System 2022, Completions Survey, https://nces.ed.gov/ipeds/use-the-data

Nosek, B. A., Smyth, F. L., Sriram, N., et al. 2009, PNAS, 106, 10593

Park, J. J., Kim, Y. K., Salazar, C., & Eagan, M. K. 2022, J. Divers. High. Educ., 15, 218

Pearson, W. 1978, The Sociology of Science, ed. J. Gaston (Hoboken, NJ: Jossey-Bass) 38

Porter, A. M. 2019, AIP Focus On (December 2019) (College Park, MD: Statistical Research Center of the American Institute of Physics) https://www.aip.org/statistics/reports/physics-phds-ten-years-later-success-factors-and-barriers-career-paths

Porter, A., White, S., & Dollison, J. 2020, 2020 Survey of the Planetary Science Workforce. (Melville, NY: AIP Statistics) https://dps.aas.org/sites/dps.aas.org/files/reports/2020/Results_from_the_2020_Survey_of_the_Planetary_Science_Workforce.pdf

Posselt, J. 2020, Equity in Science: Representation, Culture, and the Dynamics of Change in Graduate Education (Redwood City, CA: Stanford Univ. Press)

Prescod-Weinstein, C. 2020, Signs:, 45, 421

Prescod-Weinstein, C. 2021, The Disordered Cosmos: A Journey Into Dark Matter, Spacetime, and Dreams Deferred (London: Hachette UK)

Rodriguez, S. L., Perez, R. J., & Schulz, J. M. 2022, J. Divers. High. Educ., 15, 58

Rosa, K., & Mensah, F. M. 2016, PRPER, 12, 020113

 ed. Seymour, E. & Hunter A. B. (ed) 2019, Talking About Leaving Revisited: Persistence, Relocation, and Loss in Undergraduate STEM Education (Cham: Springer)

Tate, E. D., & Linn, M. C. 2005, JSEdT, 14, 483 http://www.jstor.org/stable/40186729

Traweek, S. 1988, Beamtimes and Lifetimes: The World of High Energy Physicists (Cambridge, MA: Harvard Univ. Press)

Vazire, S. 2017, Natur, 547, 7

West, S. S. 1960, IRETEM, EM-7, 54

Whitten, B. L. 1996, NWSA J., 8, 1 http://www.jstor.org/stable/4316437

Woods, P., & Walker, A. 2022, NatAs, 6, 622

An Astronomical Inclusion Revolution
Advancing Diversity, Equity, and Inclusion in Professional Astronomy and Astrophysics
Dara Norman, Tim Sacco and Dorian Russell

Chapter 2

It All Starts with Relationships: Astronomical Collaborations with Indigenous Communities

Kathryn (Kat) G Gardner-Vandy and Daniella M Scalice

2.1 Introduction

Those who hold power within colonial systems, including the majority of the Westernized STEM community, have an extraordinary opportunity to acknowledge our roles in structural racism and reexamine our unintended biases. We also have the opportunity to engage in authentic relationships and collaborative initiatives that honor the lives and cultures of the Indigenous communities that have been present since time immemorial on the lands, waters, and skies that we occupy and call our homes and workplaces. Every aspect of the global, dominant Western/Eurocentric culture remains steeped in colonialism, which by definition erases Indigeneity. Through the lens of racial and social justice, we can examine how Western cultures have interacted with Indigenous communities in the past and, looking forward, imagine how they can coexist and thrive. This view is particularly important as we strive toward ensuring the STEM community and the ways in which STEM is accomplished embody and express the full diversity of humanity.

This chapter offers some historical views of how the Westernized STEM community has operated in a system of oppression, in the absence of relationship and trust with Indigenous communities, and reimagines how institutions and scientists can, through building relationships and trust, uphold processes of accountability, equity, justice, and healing today. We offer guidelines for working with Indigenous communities that are based in the knowledge that relationship- and trust-building is the foundation upon which all collaborations should be centered. Such a relational foundation not only reflects Indigenous philosophy, but also upholds the truth that Indigenous and Western knowledge systems can co-exist, share equal importance and value, and reinforce one another.

The original form of this document was collaboratively written with many of our Indigenous colleagues and friends for the Planetary Science and Astrobiology

doi:10.1088/2514-3433/ad2174ch2　　　　2-1

Decadal Survey 2023–2032 (Gardner-Vandy et al. 2021). Much of the material in this chapter is excerpted or adapted from that collaborative work.

2.2 Historical Perspective and Context

Because this chapter focuses on Westernized STEM practitioners and educators building relationships with Indigenous communities as an emergent diversity, equity and inclusion (DEI) framework, it is pertinent to point out that a relationship between them already exists. The history of interaction between Western and Indigenous cultures is an ongoing narrative of colonialism, genocide, ethnocide, land theft, broken treaties, assimilation, erasure, extraction, and appropriation. The relationship between Western and Indigenous cultures in the United States started with the colonization of what is known currently as North America and continues today. It is characterized by violence, and it is not of the choosing of Indigenous communities.

The US is home to 574 federally-recognized, culturally-unique Indian Nations (NCAI 2022), yet mainstream American consciousness has homogenized their individual nature. Settler-colonialism established practices of extraction and appropriation of Indigenous lands, cultures, and knowledges that continue to this day. This narrative is defined by cognitive imperialism (Battiste 2017) and cultural hegemony (Cole 2020)[1] in which Western knowledges, pedagogies, and values dominate, and those of Indigenous communities are marginalized and diminished. Within research contexts, Indigenous peoples were (and continue to be) exploited and objectified for purposes of study, and their knowledges continue to be described as myths, legends, and stories. These legacies inform attitudes and practices within the current, practiced culture of Westernized STEM/education that are disrespectful and inequitable. They have resulted in lasting impacts on who is recognized as legitimate, identified as expert, and allowed to participate in the scientific enterprise broadly, and define the experience of many scholars of color today (Blanchard 2020).

Western educational systems—formal and informal—stand in a direct lineage with the US federal government's former Indian boarding school system, the truth of which has largely been written out of the history books but is thankfully coming to light now. Between 1819 and 1969, the Federal Indian boarding school system consisted of 408 Federal schools across 37 states or then-territories, including 21 schools in Alaska and 7 schools in Hawai'i (Newland 2022). They were government-funded, and often church-run. Indigenous children were forcibly abducted by government agents, sent to schools often hundreds of miles away from their homes, and, among other severe abuses for other "offenses," they were beaten and/or starved when they spoke their Indigenous languages (NNABSHC 2022). Many, many children were murdered, and gravesites on school campuses are being

---

[1] As Marie Battiste (2017) describes, "cognitive imperialism is a term that describes the mental, emotional, destructive, and traumatic effects of the experience of individuals and peoples forced to be educated and living under Eurocentric colonialism and imperialism." Cultural hegemony refers to "domination or rule maintained through ideological or cultural means", e.g., through institutions, values and norms (Cole 2020).

discovered now. Echoes of this legacy are heard in the halls of today's classrooms, curricula, and out-of-school-time programs that systematically leave out and/or tokenize Indigenous knowledges, cultural practices, pedagogies, and language instruction.

Indigenous knowledge systems contain information that is inherently, what today is described as, scientific. Indigenous knowledges and languages hold profound understandings of the nature and workings of the physical Universe and reflect sophisticated cosmologies. Indigenous technologies have been and continue to be made possible by a deep knowledge of what today are called physics, mathematics, and engineering. Indigenous communities are the experts of their lands/waters/plants/animals/skies, as reflected in millennia-long, high-resolution data sets of place, held in community and oral tradition (David-Chavez & Gavin 2018). These data sets are intersectional across many Westernized STEM disciplines, including astronomy, geography, geology, climate and environmental science, oceanography, ecology, biology, and agriculture, modeling the interdisciplinary aspirations we hold for the sciences today. Indigenous methodologies for data acquisition and curation and knowledge dissemination may differ from those of Westernized STEM, yet are no less accurate or valid.

2.3 Westernized STEM Systems and Institutions Perpetuate Exclusion

Because Westernized STEM research and education systems and institutions are rooted in colonization, capitalism, and patriarchy, we as practitioners are constantly at risk of expressing those values therein in our work. For example, in Westernized STEM educational programs intended to serve Indigenous youth, instruction typically comes from non-Indigenous educators, and in one of two ways: without any cultural component ("STEM-only") or with one that is presented secondarily ("STEM-first"). In the 'STEM-Only' paradigm, Westernized STEM knowledge, conveyed by non-Indigenous educators, is the only material being shared. Unfortunately, this is the typical situation for the teaching and learning of STEM and it reinforces the Western-dominant power dynamic that historically and contemporarily marginalizes and isolates Indigenous youth from themselves, their culture, and indeed, from pursuing careers in Westernized STEM. Perhaps more insidious is the 'STEM-First' paradigm, in which non-Native experts convey Westernized STEM concepts and then include Indigenous concepts and/or practitioners as a secondary, add-on component, minimizing it overtly and by design.

The absence of proper education on Indigenous histories, cultures, language, and knowledge systems have led to several US states mandating the inclusion of Indigenous content within school curricula (Janzer 2019; Belsha 2021). While this is a powerful step, scenarios wherein non-Indigenous educators attempt to convey Indigenous content in the absence of Indigenous voice or participation have increased. Much of the time, these efforts at best misrepresent and at worst tokenize and continue to minimize Indigenous culture (Williams 2020). In most cases, Western pedagogies prevail (Cote-Meek 2014), and the mandate lacks direction, guidance, and support toward relationship-building to enable co-creation and collaboration.

Another common practice that perpetuates exclusion happens via the grant proposal writing process. In many cases wherein Westernized STEM educators seek to collaborate with an Indigenous community, the initial approach to the community comes at the time when the grant proposal is being written. Ideas about what to propose (for example, a planetarium show that features the constellations as known to a particular community) are often fully formed by the Western institution in the absence of a relationship with the community; in other words, ideas about what to propose and why are not co-created. Similarly, the initial approach to a community made by Westernized STEM educators is often centered on the desire to bring an existing curriculum or program to Indigenous learners. These types of approaches constitute an "ask," effectively circumventing authentic relationship-building and co-creativity of ensuing activities.

The story is similar for research conducted by Westernized scientists on the lands/waters/plants/animals/skies of Indigenous communities. While many scientists are respectful in their approach to sampling during fieldwork, there are some who would act outside ethical and legal frameworks. Recent headlines have revealed acts of unethical sampling and geovandalism at a federally-protected archeological site on Indigenous lands, just a few feet from petroglyphs sacred to a California tribe, even when a permit for sampling was denied (Sahagún 2021).

More generally, Indigenous knowledges are rarely sought and typically regarded as unscientific and therefore of little value. When Indigenous knowledges are included, such as in climate studies, the vast majority (87%) practice an extractive model in which outside researchers use Indigenous knowledges with minimal participation or decision-making authority from communities who hold them (David-Chavez & Gavin 2018). Inclusion of community members in the research is typically relegated to the purveyance of services such as being a local guide or transporting gear vs. as a knowledge holder, research collaborator, or cultural authority. Occasionally, young people may be included in the research in an internship capacity, and frequently there is an offering of a lecture about the science for the community—both of which reflect imbalanced power dynamics.

Data and samples are often extracted without community input, spiritual permission, or reciprocity. Data sets typically fail to include either proper cultural provenance to the lands, waters, or skies from which they were collected, or accompanying metadata about Indigenous interests and concerns for appropriate and responsible use into the future. Publications resulting from this research rarely include acknowledgement, let alone co-authorship (Anderson & Christen 2019). When the research is completed, there is typically no ongoing relationship or connection.

Examples of exclusion can also be seen in astronomical nomenclature. When deciding to use Indigenous words as names for celestial bodies, the hubris of assumptive right of access by the Western "discoverer" to the cultural/intellectual property of the Indigenous language is starkly apparent (Tiscareno et al. 2021), even when the intention may be to honor the Indigenous culture from which the word is taken. In most cases, the word is simply appropriated, without asking permission from cultural leaders: an extractive "take" scenario. In some cases, permission is sought and granted: an "ask and take" scenario. This scenario, however, without the presence of relationship and trust between the scientists and the Indigenous community, is an "ask, take, and leave" scenario, such as with the recent naming of the Arrokoth Kuiper Belt Object (Hautaluoma et al. 2019; Tiscareno et al. 2021). In the absence of relationship and on a short timeline, these scenarios often cause harmful ripple effects within communities, for example with regard to who was or was not asked for permission, (personal communication to the authors, September 2022). Co-discovery and co-authorship are all but unheard-of.

Another example in astronomical nomenclature is the recent NameExoWorlds project of the International Astronomical Union, a powerful entity in the Westernized astronomy community, wherein "…speakers of Indigenous languages are encouraged to propose names drawn from those languages" (NameExoWorlds[2]). Via a contest construct, the official names for exoplanets were solicited from citizens across the globe and then vetted by small, majority non-Indigenous committees in each participating country, with final decisions made by votes from the citizens of each country. Issues of cultural appropriation arose when the US committee realized many of the submissions in Indigenous languages were made by non-Indigenous persons or entities. Remediative efforts were put in place, but in the absence of relationships with Indigenous communities, these issues are the norm (Tiscareno et al. 2021).

A final alarming situation within the astronomy community is the imminent construction of the contentious Thirty Meter Telescope (TMT) on Mauna Kea in Hawai'i. For several years, thousands of Native Hawaiians and Kanaka Maoli practitioners have protested the addition of this extremely large telescope to the top of the sacred volcanic summit. Proponents speak to the TMT's scientific importance, a ground-based telescope capable of probing deep into the sky, yet the project "has become a symbol of historical inequalities in Hawai'i" and perpetuated land seizure and cultural erasure (Witze 2020). Before protests gained national and international attention the TMT project had not called for or valued the local Indigenous populations' input and guidance, and protesters challenged the way Western institutions do science on the summit. As quoted in Witze (2020), anthropologist Dr Noe Noe Wong-Wilson, the leader of the Mauna Kea *kia 'i* protesters, says: "Astronomers look at us like we're the bad guys, like we're intruding on their space. It's quite the opposite: they're in our space." In their paper for the National Academy of Sciences Decadal Survey on Astronomy and Astrophysics, Kahanamoku et al. (2020) summarize three major concerning issues in the TMT conflict, including the use of law enforcement against protesters, the lack of benefit of Mauna Kea astronomy to Native Hawaiians, and the disconnect between astronomers and Native Hawaiians, and they recommend to immediately halt TMT plans and start dialoguing with the local communities. Small strides have been made recently as the Hawaiian government has mandated an advisory board

---

[2] http://www.nameexoworlds.iau.org/

with the requirement that at least two of the seats be filled by Native Hawaiian people (McAvoy 2022), yet this does not compensate for the disrespect the local people have encountered in protecting their sacred space.

2.4 Why Relationship Building?

The above ways of thinking and working are expressions of how Westernized STEM systems and institutions maintain exclusion when relationship and trust are not the priority and/or are absent. The need for relationship and trust is reflected in Indigenous metaphysics and worldviews of the Universe as a place of interconnectedness, interdependence, and relationality (Deloria 1973; Cajete 2000; Littlebear 2020). Everything flows from relationships, all are co-creators, all are responsible for one another, and reciprocity is the norm. Ideal relationships continuously honor the sovereignty and self-determination of all Indigenous communities, and celebrate each one as a unique entity. Ideal relationships embody the values of humility, gratitude, and respect at the center of Indigenous culture. Relationship and trust facilitate co-creativity and service.

In this task, motivations must be clear. There is much talk about the motivation for DEI efforts being driven by the fact that diverse teams produce more innovative results. This motivation centers Western perspective and power, and there are better motivations to be had: "'Diverse perspectives yield the best science' is a true statement, but it's one that commodifies the lived experience of marginalized people by reducing them to their contributions to productivity. It's a capitalistic framework that shirks the basic truth that cultivating a field where the norm is respecting the humanity and validity of all people is the right thing to do for no reason other than that it is right" (Ivory & Ivory 2020). No other motivations are needed. We build relationships and trust and co-create with Indigenous communities because it is the right thing to do.

2.5 Relationship Building as a DEI Framework

Our key recommendation is for agencies, scientists, and educators to engage in relationships and build trust with Indigenous communities, so research collaborations and educational programs involving Indigenous communities and their knowledge and lands/waters/plants/animals/skies can be based in co-creation and authentic service. In the absence of relationship and trust, activities risk being STEM-only or STEM-first, failing to center community needs, places, perspectives, pedagogies, and cultures, and reinforcing damaging power dynamics.

To incentivize good practices, it would be useful if agency solicitations reflect this crucial component of building relationship and trust, building funding into those activities themselves. Agencies could design a special two-part solicitation architecture that begins with a solicitation focused exclusively on investment in relationship building activities, allowing for flexibility on what those activities need to be based on the unique context of each community involved and the amount of time needed (ideally 3–5 years minimum). Activities may include informal meetings and gatherings, participation in ceremonies, sourcing community input via "Round table" dialogues or other means, meetings with tribal officials and/or Councils, tours, knowledge exchanges, etc. The second part is a solicitation for programs and collaborations when the relationship has reached a point that parties are ready to start co-creating or co-researching. This part must be flexible, as it cannot be known at the beginning of the process exactly when the need for these funds will arise. Relationships and collaborations need the flexibility and support to progress at "the speed of trust." Agencies need to be prepared to continue investing in these collaborations for the long-term.

To mobilize individual or small teams of scientists to engage in relationship building and eventual programs and collaborations with Indigenous communities, agencies could implement a small grants program for awarded investigators. Even ~$10–20K could catalyze a new relationship, and/or enable scientists to participate in and add value to the activities of other teams that are more advanced in their programs and collaborations.

To ensure co-created programs and collaborations are progressing authentically and appropriately, there should be implementation of an Office of Tribal Relations to manage and coordinate all the tribal-related work going on in the agency or parts thereof. It would be beneficial for helping to avoid duplication, enhancing connectivity, and coordinating evaluation activities as appropriate. It could administer smaller pots of funding to supplement ongoing activities and/or respond to emerging opportunities. Staff would be experienced and fluent in the dynamics of these relationships and the nuances of the work.

The Office could serve as the agency's single touch point to Indigenous communities, as current paradigms create confusion with numerous programs, offices, and personnel. It can serve the agency by ensuring their policies and practices are consistent with Indigenous communities' needs, thoughts, and desires on research and education. Most importantly, the Office's main goal would be to provide care, guidance, support, networking, and professional development to all.

As researchers, educators, and institutions become more open to developing relationships with Indigenous communities, research and training on the issues of cultural and intellectual property, Indigenous data sovereignty, and processes of knowledge management should be foundational for any program or collaboration with Indigenous communities to ensure Indigenous knowledges are not being misappropriated (Anderson & Hudson 2020). These trainings will support researchers and educators to develop a more nuanced understanding of Indigenous knowledges; acknowledge the contemporary relevance, cultural validity, and application of Indigenous knowledges; accept the differences between Indigenous and Western ways of knowing; and acknowledge the Indigenous community's authority of their knowledge.

2.6 Suggested Guidelines for Building Relationships and Trust

The following steps are intended to help guide the process of relationship building for citizens of the planetary science/astronomy/astrobiology community. It must be noted that each Indigenous community is a unique entity; the process for relationship building with one nation will not look the same with another.

2.6.1 Before Initiating a Relationship

The initiating party should endeavor to learn as much as possible about the history and culture of the Indigenous community being approached, seeking primary sources of information and practicing healthy critique of secondary sources. This includes honoring Indigenous communities' high resolution, longitudinal, cross-disciplinary data sets about their places and the cosmos we all inhabit. Sovereignty over their data must be respected (Native Nations Institute 2022). Most importantly, the initiating party must self-reflect and prepare for an ongoing effort to sustain a lasting relationship with the Indigenous community. Such partnership building efforts take time. If upon self-reflection, it is found that the initiator is not prepared for the responsibility of relationship building, co-creation, and service, then we recommend suspending the initiative and returning to the idea only when such a commitment is feasible.

2.6.2 Initiating a Relationship

Initiating a relationship with an Indigenous community must be guided solely by the desire to build that relationship toward co-creativity and service. Understanding the importance of Elders in the community is essential to understanding the community as a whole. Initial communications should not carry an "ask," such as joining an event you are hosting and offering a performance, cultural activity, or blessing. Such an ask is extractive, disrespectful, and offensive if not done in the context of relationship and trust. Do not approach an Indigenous community with your research project or educational program set. Ensure your ideas percolate with those of your Indigenous partners. Let go of your need to be the sole expert. Prioritize the needs and vision being articulated at a community level. Begin with an introduction, and clearly articulate a desire to truly know one another and one day co-create and serve the community. If such an invitation is received and supported by the Indigenous community, begin the process of getting to know one another. If invited, physically visit the community in their space, bringing your whole self to such meetings, ready to communicate with and learn from each other.

Gift-giving is an important gesture when visiting a community, so consider bringing a small token of your appreciation when invited into their space. If a meeting occurs at an institution instead of on Indigenous land, the host should still offer a gift, and celebrate the gathering by providing food. Listen more than you speak. Resist any urge to control or drive the conversation. Be humble in your role as guest and learner and offer gratitude for the opportunity to receive knowledge. Finally, always follow up with notes of gratitude, recognizing that the Indigenous community is under no obligation to commit to a relationship with you.

2.6.3 Building Relationships

Once a relationship is started, be prepared to spend the time necessary to cultivate the relationship consistently and indefinitely. Create and hold space for the community to lead and direct, not the other way around. Prioritize the

Indigenous community as a source of credible information. Ensure the needs of the community shape the shared vision for the collaboration. Be flexible if the needs of the community change. Trust is the key. If trust is not present or is lost, the relationship is lost. If you say you're going to do something, do it.

Share of yourself and make a habit of sending notes/emails and making calls to the community to ask how people are doing. Show up to the community if there is an emergency and/or help organize a local response. If you become aware of a community-wide or public event, make sure to attend, and while doing so, be willing to engage in the community fully. Ensure the relationship with the community is not simply a scientific endeavor for your institution.

2.6.4 Ensuring Lasting Relationships

Together with your Indigenous partners, it may be required to seek blessings and permission from tribal leadership (Tribal Council or other governing body) to continue the relationship. Doing so may result in Resolutions of Support, Memoranda of Understanding, and/or official commitment of resources. Follow your partners' leadership in how to move forward in this process, making sure the focus is not on an "ask." Ensure the community/tribe is properly represented in the grants you write together (Co-PI, Co-I, etc.). Budget to compensate individuals/ groups for their expertise and contributions at the same rate as non-tribal consultants. Ensure relationships and contributions are properly identified and visible in publications, reports, and other research outputs. This might include co-authorship or acknowledgements, and/or using tools like the Traditional Knowledge and Biocultural Notices (Anderson & Hudson 2020) which clarify Indigenous interests in the research, data, and collected traditional knowledge. Transparency and integrity in collaborations is central to building trust for future collaborations and partnerships.

It is imperative for institutions and individuals to understand that relationships with Indigenous communities be built with long-term effort in mind. If plans change from the institution or if there is a job change or retirement, plans need to be made in advance to ensure that service to the Indigenous community is not halted. A productive way to ensure a lasting relationship is to include many members of your institution/scientific community in the relationship. This will safeguard the relationship that was built and maintain trust.

2.7 What's Possible in the Presence of Relationship?

Relationship and trust enable the potential for authentic partnership and co-creativity in STEM research and education. It enables a space in which to learn about the priorities the Indigenous community has for their youth, lands, waters, and skies, and ensure those priorities guide how the collaboration will unfold.

For the collaborations we and many of our colleagues are involved in, the priorities are to bring together Indigenous and Westernized scientific knowledge into educational resources and programs for Indigenous youth, and we have learned that it matters significantly how it is done. Reframing STEM education for Indigenous youth from a conversation about their achievement and survival in a Western world to one about the vitality of community and sustainability requires that we actively reconfigure what counts as science learning and who is teaching it (Bang et al. 2009). In co-creating place-based educational opportunities for Indigenous youth, inter-weaving cultural traditions, arts (STEM-to-STEAM), language, and community partnerships is key in order to provide authentic service (Lopez et al. 2013; Walkingstick & Bloom 2013). The integration of Indigenous languages in learning environments is imperative for Indigenous youth's sense of identity and academic success (Reyhner 2017). Representation and role modeling by Indigenous leaders and educators has significant effects on Indigenous learners' sense of belonging (Covarrubias & Fryberg 2015). Indigenous educators and cultural practitioners are critical for facilitating Native youth's curiosity and engagement with STEM components.

There are two pedagogies that offer educators the possibility to do these things and more. One is Two-Eyed Seeing (TES) and the other is Culturally-Based Dual-Learning (CBDL), and the foundation of both is respect for Indigenous sovereignty and self-determination. TES leverages the strengths of both Indigenous knowledges and Westernized STEM, where the learning of both occurs in tandem in a non-competing way throughout the curriculum (Hatcher et al. 2009; Gardner-Vandy et al. 2022). With CBDL, Indigenous concepts and epistemologies of the local community are presented first—always and only by Native Elders, leaders, and experts, and then corresponding Westernized STEM concepts and epistemologies are shared by experts in those fields. With both TES and CBDL, the learning environment is cultivated from the foundational understanding that the two bodies of knowledge and ways of knowing reflect, reinforce, and resonate with each other, bringing Indigenous knowledge and Westernized STEM together into a relationship wherein no one is dominant over the other.

These approaches center, privilege, and make primary Indigenous knowledges and community learning priorities. They go beyond STEM-First and STEM-Only approaches, and provide a way for Indigenous youth to approach Westernized STEM from an asset-based (California Department of Education 2023) position of strength, grounded in their own cultural knowledge of Westernized STEM. In these ways, the community's knowledge is the philosophical framework, foundation, and lens through which Westernized STEM concepts are viewed and learned. Indigenous learners then become grounded in Indigenous science and in the understanding that their culture is inherently scientific and, as part of that culture, they can see themselves as scientists. They understand that learning Westernized science can flow naturally for them and that they, too, may want to pursue becoming an Indigenous scientist. This has lasting implications for the diversity of the STEM workforce, and more importantly, for Indigenous youth and communities towards healthy social, cultural, and economic lifeways.

Wonderfully, there are numerous examples that demonstrate how this balance may be achieved, including but not limited to: the Loololma Model (Gilbert et al. 2011); the ašiihkiwi neehi kiišikwi myaamionki curriculum (McCoy et al. 2011); the NASA and the Navajo Nation Partnership (Barney-Nez et al. 2016; Indian Country

Today 2016; Bartels 2019); Haida Geoscience Curriculum (Smythe 2019); Native Skywatchers (Lee 2020); Arctic and Earth SIGNs (Sparrow 2020); Sharing the Skies (Maryboy & Begay 2020), 'Imaloa Astronomy Center (Kimura 2020), Native Earth | Native Sky ("NENS" 2024), the Wabanaki Youth in Science Program ("WaYS[3]"), and many, many more.

Within research, a relationship that stands out as exemplary is between the Atacama Large Millimeter/submillimeter Array (ALMA) observatory and the local Indigenous communities in San Pedro de Atacama, Chile. The ALMA community began working with the local communities well before the construction of the observatory, respectfully seeking permission from the Toconao community before starting construction even when they already had consent from the state to proceed with construction. They also conducted a study of the impact of the observatory on the environment before construction began. Once in operation, the ALMA community has continued engagement with the local community by doing activities with them and supporting the local schools. ALMA sustains a Culture Committee that works with the local people of all ages to honor their Star Knowledge and record stories of their Elders. They co-created "El Universo de Nuestros Abuelos: The Universe of Our Elders," a publication that shares the vision of interweaving two "remarkable" projects: celebrating the traditional ways of knowing of the traditional inhabitants of the Atacama Salt Flat and the ALMA facility itself ("El Universo" 2016). Valeria Foncea, ALMA's Education and Public Outreach Manager, states ALMA's philosophy succinctly and beautifully: "Whatever we do, we ask them first" (personal communication, August 16, 2022).

A model on what is possible in the presence of relationship is found in the Kūlana Noi'i guidelines for achieving reciprocal, community-research collaborations (Kūlana Noi'i Working Group 2021). They are intended as a starting point for deeper conversation, presenting a set of ideas, values, and behaviors that when applied alongside hard work, can build more just and generative relationships between researchers and community. The Kūlana Noi'i reflect the perspectives and responsibilities of both community members and researchers entering into a partnership together, and support equitable and sustained relationships to form where the interests of research and community overlap. The kūlana are grouped into two sections. First, *Building and Nurturing Pilina (Relationships)* which outline practices for establishing a foundation of strong and equitable partnership. Many of these kūlana can (and should) be practiced before a research question emerges and a proposal is developed. And *A'o aku, a'o mai/Aloha aku, aloha mai (Knowledge given, knowledge received/Love given, love received)* which include practices and questions to be considered once a community–researcher relationship has been established. These kūlana can help guide a collaborative research efforts to achieve more impactful outcomes that benefit all partners in the long-term.

---

[3] http://www.wabanakiyouthinscience.org

2.8 Conclusion

The great Dr Robin Wall Kimmerer offers metaphors of autonomy and co-existence between the two knowledge systems that compromise the integrity of neither (Kimmerer 2020). With the "Three Sisters Garden," she describes a system of knowledge mutualism, with Indigenous, Elder knowledge (the corn) acting as the intellectual scaffold which in turn guides the scientific knowledge (the beans) which enrich the symbiosis. The relationship (the squash) creates a climate for multiple species of knowledge to grow. Crucially, she adds a fourth sister–*us*–to tend the whole garden. So much is possible when relationships are cultivated from the rich soils of respect for sovereignty and self-determination. Co-creativity and service grow when the knowledge and wisdom resident in both Indigenous and Westernized STEM are brought together into partnerships and collaborations. Working in this way opens up pathways to reconciliation (Littlebear 2020) and restoration, and the resulting work will indeed be made whole. None of this is possible without the foundation of relationship and trust.

References

Anderson, J., & Christen, K. 2019, J. Radic. Librariansh., 5, 113

Anderson, J., & Hudson, M. 2020, The TK and biocultural (BC) labels initiative: overview (A Presentation to the NASA AI/AN Working Group) https://www.youtube.com/watch?v=z3QTNEP8T64

Bang, M., Medin, D., & Cajete, G. 2009, University of Washington's Science Education, 12, 1

Barney-Ncz, A., Carron, A., & Scalice, D. 2016, NASA and the Navajo Nation, A Presentation to the NASA AI/AN Working Group, http://astrobiology.nasa.gov/education/nasa-and-the-navajo-nation/

Bartels, M. 2019, NASA and Navajo Nation Partner in Understanding the Universe. Space.com

Battiste, M. 2017, ed. M. A. Peters Encyclopedia of Educational Philosophy and Theory (Singapore: Springer)

Belsha, K. 2021, 'They'll know more than I ever knew': More states move to require lessons on Native American history and culture. Chalkbeat.com, https://www.chalkbeat.org/2021/8/4/22607758/states-require-native-american-history-culture-curriculum

Blanchard, P. 2020, Centering Native Voices Within Earth Sciences: An Inquiry into Opportunities and Challenges Experienced by Native Students, Early-Career Scholars and Scientists. Earth's Future (Washington, DC: American Geophysical Union)

Cajete, G. 2000, Native Science: Natural Laws of Interdependence (Santa Fe, NM: Clear Light Publishers)

California Department of Education 2023, Asset-based pedagogies. https://www.cde.ca.gov/pd/ee/assetbasedpedagogies.asp

Cole, N. 2020, What is cultural hegemony? ThoughtCo.com. https://www.thoughtco.com/cultural-hegemony-3026121

Cote-Meek, S. 2014, Colonized Classrooms: Racism, Trauma, and Resistance in Post-Secondary Education (Nova Scotia: Fernwood Publishing)

Covarrubias, R., & Fryberg, S. A. 2015, Cultur. Divers. Ethnic Minor. Psychol, 21, 10

David-Chavez, D., & Gavin, M. 2018, ERL, 13, 123005

Deloria, V. 1973, God Is Red: A Native View of Religion (Ann Arbor, MI: Fulcrum Publishing)

El Universo de Nuestros Abuelos: The Universe of Our Elders 2016, https://almaobservatory.org/wp-content/uploads/2016/11/alma-etno_2013.pdf

Gardner-Vandy, K., Scalice, D., Chavez, J. C., David-Chavez, D. M., Daniel, K. J., et al. 2021, Bull. AAS, 53, 471

Gardner-Vandy, K., Utley, J., Just, A., Hathcock, S., & Stansberry, S. 2022, Rev. Educ. Res., (submitted)

Gilbert, W. S. 2011, ed. J. Reyhner, W. S. Gilbert, & L. Lockard Honoring Our Heritage: Culturally Appropriate Approaches to Indigenous Education (Flagstaff, AZ: Northern Arizona Univ.) 43

Hatcher, A., Bartlett, C., Marshall, M., & Marshall, A. 2009, Green Teach., 28, 3

Hautaluoma, G., Johnson, A., & Buckley, M. 2019, Far, far away in the sky: New Horizons Kuiper Belt flyby object officially named 'Arrokoth', https://www.nasa.gov/feature/far-far-away-in-the-sky-new-horizons-kuiper-belt-flyby-object-officially-named-arrokoth

Indian Country Today 2016, Weaving Diné Knowledge with NASA Science for Community Education.

Ivory, K. 2020, #BlackInAstro Experiences: KeShawn Ivory. An Invited Contribution to Astrobites: The ASTRO-PH Reader's Digest http://astrobites.org/2020/06/19/black-in-astro-keshawn-ivory/

Janzer, C. 2019, U.S. News and World Report., https://www.usnews.com/news/best-states/articles/2019-11-29/states-move-to-add-native-american-history-to-education-curriculum

Kahanamoku, S. S., et al. 2020, National Academy of Science Astro2020 Decadal,

Kimmerer, R. 2020, The Fortress, The River, and the Garden: A Model for Integration of Science and Indigenous Knowledge. Indigenous Education Institute, Webinar Series.

Kimura, K. 2020, Imiloa: Sharing Hawai'i's Legacy of Exploration, http://www.hawaii.edu/news/2019/08/30/capitol-connection-imiloa-astronomy-center/

Kūlana Noiʻi Working Group 2021, Kūlana Noiʻi v.2. University of Hawai'i Sea Grant College Program, Honolulu, Hawai'i, http://seagrant.soest.hawaii.edu/kulana-noii/

Lee, A. 2020, Native Skywatchers, http://www.nativeskywatchers.com/

Little Bear, L. 2020, Rethinking our science: blackfoot metaphysics waiting in the wings, Indigenous Education Institute's Webinar Series, http://www.youtube.com/watch?v=o_txPA8CiA4

Lopez, F. A., Heilig, J. V., & Schram, J. 2013, Am. J. Educ., 119, 513

Maryboy, N., & Begay, D. 2020, Sharing the Skies (Friday Harbor, WA: Indigenous Education Institute)

McAvoy, A. 2022, Hawaii Legislation Would Reform Management of Mauna Kea (New York: Associated Press) https://apnews.com/article/science-hawaii-astronomy-mauna-kea-8bf88d70d40cf88a16ed2e597cb27253

McCoy, T., Ironstrack, G., Baldwin, D., Strack, A. J., & Olm, W. 2011, *Ašiihkiwi neehi kiišikwi myaamionki* (Earth and Sky: The Place of the Myaamiaki). Miami Tribe of Oklahoma.

National Congress of American Indians 2022, Tribal Nations & the United States: An Introduction, https://archive.ncai.org/about-tribes

NNABSHC (National Native American Boarding School Healing Coalition) 2022, US Indian Boarding School History.

Native Earth | Native Sky 2024, https://education.okstate.edu/research/centers/native-earth-native-sky/index.html

Native Nations Institute 2022, https://nni.arizona.edu

Newland, B. 2022, Federal Indian Boarding School Initiative Investigative Report, (Office of Indian Affairs, US Department of the Interior) https://www.bia.gov/sites/default/files/dup/inline-files/bsi_investigative_report_may_2022_508.pdf

Reyhner, J. 2017, Cogent Educ., 4, 1340081

Sahagún, L. 2021, Los Angeles Times, https://www.latimes.com/environment/story/2021-07-25/horrified-professor-apologizes-for-damaging-a-sacred-site

Smythe, W. 2019, Fostering indigenous K-12 geoscience leaders: community-centered geoscience experiences and curriculum, Presentation to the NASA AI/AN Working Group, http://www.youtube.com/watch?v=pqXC2C71fN8

Sparrow, E. 2020, Engaging Learners in STEM using GLOBE & NASA Assets, https://sites.google.com/alaska.edu/arcticandearthsigns/

Tiscareno, M., Scalice, D. M., Thompson, M. L., Noviello, J. L., White, V., et al. 2021, Bull. AAS, 53, 462

Walkingstick, J., & Bloom, L. A. 2013, J. Curric. Instr., 7, 55

Williams, J. 2020, Gwaayaksichikweyan~Making Things Right (A Presentation to the NASA AI/AN Working Group) https://www.youtube.com/watch?v=TOp0kgSwcKo

Witze, A. 2020, Nat. News, 577, 457

An Astronomical Inclusion Revolution
Advancing Diversity, Equity, and Inclusion in Professional Astronomy and Astrophysics
Dara Norman, Tim Sacco and Dorian Russell

Chapter 3

An Accessible Future

A J Link

What does it mean when we make the claim that the future should be accessible? How far into the future do we have to travel before we live in an accessible world? What does a truly accessible world look and sound and feel like? What will the experiences in an accessible future be? And how will we know that the world is accessible?

As someone working on Space Law and Outer Space communication, these are questions that I think about daily. In my work, I try to center accessibility and the needs of folks who do not have access to the spaces and places where so many incredible things are happening within the space industry and the broader space community. My background is in disability rights policy work and disability justice organizing through various disability-led nonprofits, however the idea of access and creating an Accessible Future goes so far beyond disability inclusion as part of our diversity initiatives, and requires a radical reimagining of the futures we want to create.

3.1 What Is Disability?

The first part of developing a culture of accessibility is understanding where the access movement comes from and how it has developed over time. The push for access can be found in the history of the disability rights movement, and the later disability justice movement. In the United States, these movements have enjoyed a large amount of success getting society to acknowledge the barriers to access that Disabled folks experience when trying to participate in society.

But what exactly is a disability? And what exactly does it mean to be disabled? It can be hard to give a single definition for the term "disabled," because there are so many different definitions and experiences of being disabled. Having a disability is complex and complicated, and does not necessarily mean that an individual will want to identify as being disabled. Some people subscribe to a dis/ability or (dis) ability paradigm, while others do not. There are so many ongoing conversations

doi:10.1088/2514-3433/ad2174ch3

among academic, legal, and activist circles about what it means to be capital D "Disabled" and what it means to be a disabled person or a person with disability. So where do we start?

An internet search for "What is a disability" will get lots of different answers. There are multiple different dictionary definitions,[1] there are international organizational definitions (e.g. World Health Organization definition), and there are even legal definitions. In the US, Section 504 of the Rehabilitation Act (Section 504) of 1973 and the Americans with Disabilities Act of 1990 along with ADA Amendments Act of 2008 all provided legal definitions for having a disability. Legal definitions of disability are important for shaping the accommodations process, but they aren't the only way of understanding disability. There are lots of models of disability. For instance, the definition provided in the *Convention on the Rights of Persons with Disabilities (CRPD)*, a United Treaty published in 2007, is a helpful one to use:

> Persons with disabilities include those who have long-term physical, mental, intellectual or sensory impairments which in interaction with various barriers may hinder their full and effective participation in society on an equal basis with others.

This definition is particularly useful because it reflects the human rights affirmation models of disability, which includes the medical model of disability ("impairments"), the social model of disability ("interaction with various barriers may hinder their full and effective participation in society"), and the legal model of disability ("equal basis with others") to produce a framework that acknowledges that, while disabilities can be limiting, it is often *barriers to access* that cause the actual disabling experience. There are also other models of disability, such as the charity model of disability, that infantilizes disabled people and often removes their agency, or the diversity model of disability, that focuses on embracing Disability as an identity similar to other socially constructed identities like race, sexual orientation and gender identity (SOGI), nationality, and so on. I prefer the definition provided by the Convention on the Rights of Persons with Disabilities over these other definitions because it acknowledges that disability occurs as interaction between different bodyminds[2] (Price 2015) and the barriers to access that prevent them from meaningfully participating in society.

There are also subsidiary movements within the disability rights and disability justice movements, like the neurodiversity and neuroexpansive movements. Neurodiversity is a term that was popularized by Australian sociologist, Judy Singer, to describe the neurological differences that exist as part of the diversity of human experiences. Neurodiversity is the umbrella under which neurotypical minds,

---

[1] https://www.merriam-webster.com/dictionary/disabled; https://www.dictionary.com/browse/disabled

[2] Margaret Price writes that Bodymind is a term she picked up "…while reading in trauma studies (Rothschild 2000). According to this approach, because mental and physical processes not only affect each other but also give rise to each other—that is, because they tend to act as one, even though they are conventionally understood as two—it makes more sense to refer to them together, in a single term."

or those deemed "normative", and neurodivergent minds exist. Neurodivergent minds can include those labeled with Dyspraxia, Dyslexia, Attention Deficit Hyperactivity Disorder, Dyscalculia, Autism Spectrum Disorder, Tourette Syndrome, and others. While neurodiversity has gained traction in some disabled advocacy spaces, it remains more of a social term than one that has been written into legal protections. This may be due to the fact that there is not a consensus among neurodivergent individuals and neurodiversity advocates about whether self-identified neurodivergent individuals should be considered disabled, or if they even want to self-identify as disabled.

Neuroexpansive is a term coined specifically for use by Black Disabled folks as a "rejection of the term 'neurodivergent' and the ideology that undergirds it." (Ngwagwa 2022) The conception of neuroexpansiveness is an example of Dis/ability Critical Race Studies, or *DisCrit*, the intersection of Critical Race Theory (CRT) and Disability Studies (DS) that forms a framework focused on a dual analysis of race and ability (Annamma et al. 2013), analysis pushing back against a manifestation of ablenoir. Ablenoir is an intersectional term that has recently been used to describe the ableism and racism experienced by Black folks, it parallels the term "misogynoir" (which combines misogyny and noir, the French word for Black) coined by Moya Bailey to describe the intersectional misogyny experienced by Black women.

3.2 Ableism and Disabiliphobia

Now that we are grounded in the concept of disability, we can acknowledge that ableism and disabiliphobia perpetuate disabled exclusion. Ableism, as defined by TL Lewis (2022), is:

> A system of assigning value to people's bodies and minds based on societally constructed ideas of normalcy, productivity, desirability, intelligence, excellence, and fitness. These constructed ideas are deeply rooted in eugenics, anti-Blackness, misogyny, colonialism, imperialism, and capitalism. This systemic oppression that leads to people and society determining people's value based on their culture, age, language, appearance, religion, birth or living place, "health/wellness", and/or their ability to satisfactorily re/produce, "excel" and "behave." You do not have to be disabled to experience ableism.

While ableism is the system of oppression that disabled people are forced to navigate, it is the disabiliphobia that disabled people encounter that is their most consistent barrier to access. Disabiliphobia can be described as the fear of or discomfort with disability and disabled people, a fear of what disabled people represent, or a fear of potentially becoming disabled or "catching" a disability. The most common examples of disabiliphobia can be found in media portrayals of villains who are disabled, especially villains with facial differences and/or becoming disabled as part of their origin story. In the workplace, disabiliphobia manifests itself in hiring practices—specifically when recruiters and interviews make assumptions about disabled and neurodivergent candidates, based on their disabilities or appearances and behaviors that are related to their disabilities. These negative assumptions (e.g., a lack of eye contact by a neurodivergent job candidate means they aren't professional) can often prevent disabled people from entering the workforce.

3.3 From Accommodations to Accessibility

The work of disabled inclusion requires the dismantling of the systemic barriers to access that exist due to ableism and disabiliphobia. This can come in the form of individual accommodations and/or general accessibility. Accommodations can be formal or informal. The CRPD says that "reasonable accommodation" means:

> Necessary and appropriate modification and adjustments not imposing a disproportionate or undue burden, where needed in a particular case, to ensure to persons with disabilities the enjoyment or exercise on an equal basis with others of all human rights and fundamental freedoms.

In professional settings in the US, the ADA defines "reasonable accommodations" as:

> A "change or adjustment to a job or work environment that permits a qualified applicant or employee with a disability to participate in the job application process, to perform the essential functions of a job, or to enjoy benefits and privileges of employment equal to those enjoyed by employees without disabilities."

These accommodations are based on individual access needs, and usually must be requested by people with disabilities. This places the responsibility on the disabled person to explain the different barriers to access they are experiencing, which could force someone into unnecessary or unwanted disclosure. This is especially difficult in professional settings where stigmas against disabilities and being disabled are still commonplace. A person may not want to go through the formal process of requesting reasonable accommodations, so they may opt to request informal accommodations or even informally self-accommodate in ways that may add to their work burden.

Conversely, general accessibility builds on the principles of universal design. Again, from the CRPD:

> The design of products, environments, programmes and services to be usable by all people, to the greatest extent possible, without the need for adaptation or specialized design. "Universal design" shall not exclude assistive devices for particular groups of persons with disabilities where this is needed.

Accessibility, in the context of human rights for people with disabilities as expressed in the CRPD, can simply be understood as the right of access or the right to access spaces without barriers. Accessibility in this context is collective. Accessibility is focused on making places as accessible as possible for as many people as possible, including those who have not yet interacted with an environment. This understanding of accessibility highlights two of the *10 Principles of Disability Justice as conceptualized by Sins Invalid*: Interdependence and Collective Access. There are other principles of Disability Justice that are useful when developing a deeper understanding of what access needs are and what radical accessibility means, that is, Recognizing Wholeness and Intersectionality.

We can take these lessons and move beyond a disability-focused conception of access to a multidimensional conception of accessibility that is aware of multiple barriers to access based on the multilayered identities and experiences of all people. This is what it means to move accommodations to universal design to broader accessibility.

3.4 Why Access Is Important for Space Science

What does all of this mean for the space sciences community? Why should the space community care if we're accessible or inaccessible? Because space is for everyone, as we are told constantly, by *individuals*, the *government*, and *even at the level of the United Nations*[3]. Yet, space has not been and is not for everyone. We have erected so many barriers to entering our community and its professions and we have done it at the earliest opportunities along academic and career pathways. We have excluded so many people, unintentionally and intentionally.

The topic of ableism in academia has been explored in several places including articles in the *National Library of Medicine, Nature*, and in a full volume book aptly titled *"Ableism in Academia."* (Brown & Leigh 2020; Brown & Ramlackhan 2022; Powell 2021). The experiences of disabled folks in higher education are similar across academic disciplines. Disabled folks face barriers throughout their academic career: from the lack of accommodations available for high stakes tests like the SAT, ACT, MCAT, GRE, LSAT, etc, to professors not creating accessible learning environments with things like flexible attendance policies and readily available class recordings, to labs that do not incorporate universal design principles so disabled people can navigate them. Academics should make more of an effort to collaborate with groups like *Disabled in Higher Ed* and the *National Disabled Law Students Association* who have worked to make higher education spaces more accessible for disabled folks. In addition to working with disabled people, there are resources for *universal design in employment* and *information on inclusive design in higher education*. The space community also has the added resources of *AstroAccess*. AstroAccess is a project dedicated to promoting disability inclusion in the space sector, including in STEM fields.

We also must recognize that disabled people have unique expertise that we need if we are to exist, as a species, in space. Outer space is still an unknown environment, but we do know that it is harsh and potentially disabling. As Dr Sheri Wells-Jensen

---

[3] U.N. Office for Outer Space Affairs Access to Space for All https://www.unoosa.org/oosa/en/ourwork/access2space4all/index.html

(2018) has argued, *some situations in space may be better suited for disabled astronaut.* We need the knowledge and expertise of disabled people, along with everyone else, to be as prepared as possible for all the unknowns in humanity's future beyond Earth.

References

Annamma, S. A., Connor, D., & Ferri, B. Race Ethn. Educ, 16, 2013, 1

Brown, N., & Leigh, J. Ableism in Academia: Theorising Experiences of Disabilities and Chronic Illnesses in Higher Education (London: UCL Press) 2020,

Brown, N., & Ramlackhan, K. High. Educ., 83, 2022, 1225

Cook, T. M. Temp. LR, 64, 1991, 393

Invalid, S. 2015, 10 Principles of disability https://www.sinsinvalid.org/blog/10-principles-of-disability-justice

Lewis, T. 2022, Working definition of ablism update https://www.talilalewis.com/blog/working-definition-of-ableism-january-2022-update

Ngwagwa, 2022, Neuroexpansive™ Thoughts, *Medium* https://medium.com/@ngwagwa/neuro-expansive-thoughts-9db1e566d361

Powell, K. 2021, Natur, 598, 221

Price, M. 2015, Hypatia, 30, 268

Rothschild, B. 2000, The body remembers: The psychophysiology of trauma and trauma treatment, (New York: W. W. Norton & Company)

The Planetary Society 2019, Space for Everyone: Strategic Framework 2019–2024,

Wells-Jensen, S. 2018, The Case for Disabled Astronauts, Scientific American Newsletter https://blogs.scientificamerican.com/observations/the-case-for-disabled-astronauts/

An Astronomical Inclusion Revolution
Advancing Diversity, Equity, and Inclusion in Professional Astronomy and Astrophysics
Dara Norman, Tim Sacco and Dorian Russell

Chapter 4

Are We Missing Out?

Rachel Ivie and Susan White

4.1 The Relationships Among Diversity, Equity, and Inclusion

Research has shown the importance of diversity in the success of an organization and team (Campbell et al. 2013). In its *Vision for Science Excellence*, the National Aeronautics and Space Administration (NASA) notes that "diversity is a key driver of innovation, and more diverse organizations are more innovative."[1] Diversity at the leadership level is important; a lack of diversity in the leadership of an organization hampers innovation, prohibits people in marginalized groups from being recognized for their contributions, and leads to unmet expectations (Rock et al. 2016). However, diversity cannot be the only goal. Without planned efforts to increase equity and inclusion, the goal of diversity will not be achieved (Puritty et al. 2017).

To understand how to direct efforts to bring about change, we must understand the differences among diversity, equity, and inclusion. Diversity refers to the representation of various types of people in a team, organization, or society. Diversity means that people of various racial/ethnic backgrounds, genders, disabilities, sexualities, career stages, life experiences, etc., are present in a group or organization. In science, there are groups that historically have been underrepresented and marginalized. In our context, marginalization refers to the process of systematically limiting scientists' access to opportunities and resources based on one or more of their identities or background characteristics. The end result is inequity. In addition, people from marginalized groups usually face a lack of inclusion. The lack of equity and inclusion presents barriers to recruitment and retention of people from these groups. If astronomical workplaces and educational settings continue to replicate the lack of equity and inclusion present in the larger society, these settings will continue to lack diversity.

---

[1] *Explore: Science 2020–2024, A Vision For Science Excellence*, https://science.nasa.gov/science-red/s3fs-public/atoms/files/2020-2024_Science.pdf, 17

doi:10.1088/2514-3433/ad2174ch4

Two recent sets of recommendations from national advisory panels note the importance of data in efforts to increase diversity, equity, and inclusion in astronomy. The report of the Task Force on Diversity and Inclusion in Astronomy Graduate Education includes a set of recommendations around using data to measure progress toward diversity, equity, and inclusion in departments and in the larger field. The report states that a lack of data prevents documentation of progress and hinders understanding of which practices are effective:

> *The absence of a coordinated data collection effort with standard metrics has prevented departments from making meaningful comparisons—both with prior versions of themselves (to benchmark their progress) and with other astronomy departments (to gauge their equity and inclusiveness relative to that of the field and/or peer departments).[2]*

In addition, the State of the Profession Panel for the National Academies' decadal survey, *Astro 2020*, concludes that "data are key to identifying promising practices, measuring progress, and holding agencies and institutions accountable to equity-advancing values."[3] It is clear that data are necessary to assess the state of diversity, equity, and inclusion in astronomy and to measure progress toward these values. In this paper we will examine measures of diversity in astronomy by using data on the representation of historically under-represented and marginalized groups of people in the field. We will also present data on the lack of equity and inclusion in astronomy for people who are in marginalized groups. Data on equity and inclusion are just as necessary as data on representation. If data can pinpoint the areas of inequity and exclusion, the astronomy community can address these. If remedies are applied and equity and inclusion increase, diversity in astronomy should likewise increase.

In this article, we use several data sources to describe the state of diversity, equity, and inclusion in astronomy. The statistics on diversity for degree earners come either from the federal government[4] or from surveys that the American Institute of Physics (AIP) conducts annually of degree-granting astronomy and physics departments in the US. Statistics on representation among physics and astronomy faculty members come either from AIP's Faculty Member Survey or from AIP's Academic Workforce Survey.[5] Statistics on diversity among US members of the American Astronomical Society (AAS) come from the AAS Demographics Committee's Workforce Survey,[6] conducted every two to three years by the committee to document the demographics of the astronomical workforce. Data on equity and inclusion in astronomy come from

---

[2] *Final Report of the 2018 AAS Task Force on Diversity and Inclusion in Astronomy Graduate Education*, 2019. https://aas.org/sites/default/files/2019-09/aas_diversity_inclusion_tf_final_report_baas.pdf

[3] National Academies of Sciences, Engineering, and Medicine. 2021, *Pathways to Discovery in Astronomy and Astrophysics for the 2020s* (Washington, DC: The National Academies Press). https://doi.org/10.17226/26141.

[4] National Center for Education Statistics (NCES) and the National Science Foundations's National Center for Science and Engineering Statistics (NCSES)

[5] These surveys are conducted regularly by the Statistical Research Center at the American Institute of Physics.

[6] See https://aas.org/comms/demographics-committee for more information on this study.

the Longitudinal Survey of Astronomy Graduate Students, conducted by AIP and AAS, with support from the Astronomy Directorate of NSF.[7] Additional data come from the National Center for Education Statistics[8] (NCES) and the National Center for Science and Engineering Statistics[9] (NCSES).

4.2 Diversity in Astronomy

In the US in 2019, 730 people earned bachelor's degrees in astronomy and astrophysics.[10] In 2020, 305 people earned PhDs in astronomy.[11] We begin our examination of diversity in representation by examining the number and proportion of bachelor's degrees and PhDs in astronomy that women earn. Then we will show the representation of people by race/ethnicity, disability, and sexual orientation.

4.2.1 Gender

Figures 4.1 and 4.2 show an overall increase in the number of women earning bachelors and PhDs in astronomy over the last fifty years. In Figure 4.3, we see that the proportion of astronomy degrees earned by women has somewhat leveled off in the last decade. This is not a reflection of fewer women earning degrees in the field; rather, it reflects a similar growth in the number of men earning degrees in astronomy.

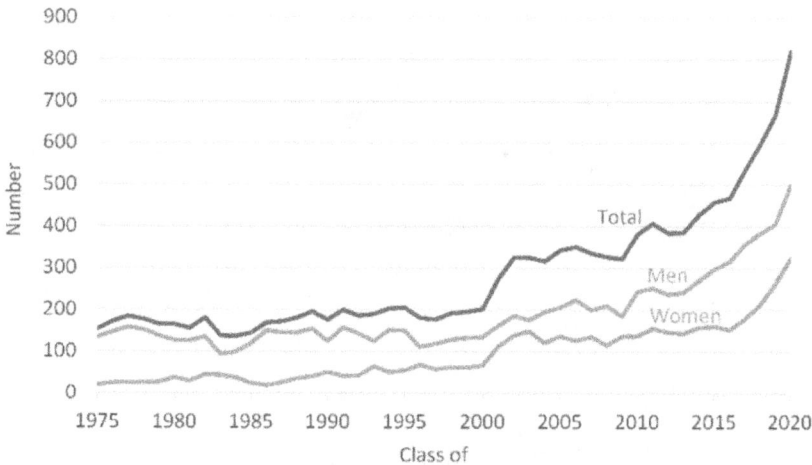

Figure 4.1. Number of bachelor's degrees earned in Astronomy, 1975–2020. Source: AIP Survey of Enrollments and Degrees.

---

[7] https://www.aip.org/content/longitudinal-study-astronomy-graduate-students

[8] See https://nces.ed.gov/ipeds

[9] See https://ncses.nsf.gov/

[10] Statistical Research Center, AIP. 2020, *Data on People from Underrepresented Groups in the Physical Sciences and Engineering, from National Center for Education Statistics.* https://www.aip.org/statistics/resources/data-underrepresented-groups-physical-sciences-and-engineering

[11] National Center for Science and Engineering Statistics, Survey of Earned Doctorates.

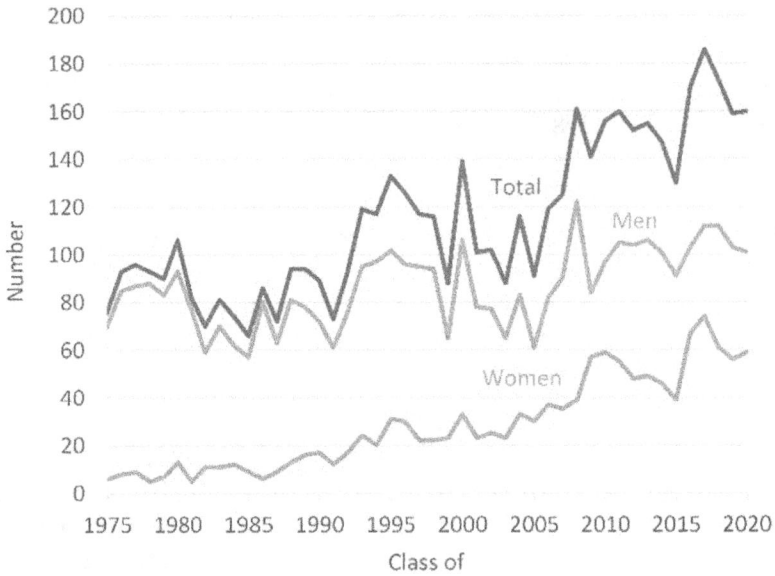

Figure 4.2. Number of PhDs earned in Astronomy, 1975–2020. Source: AIP Survey of Enrollments and Degrees.

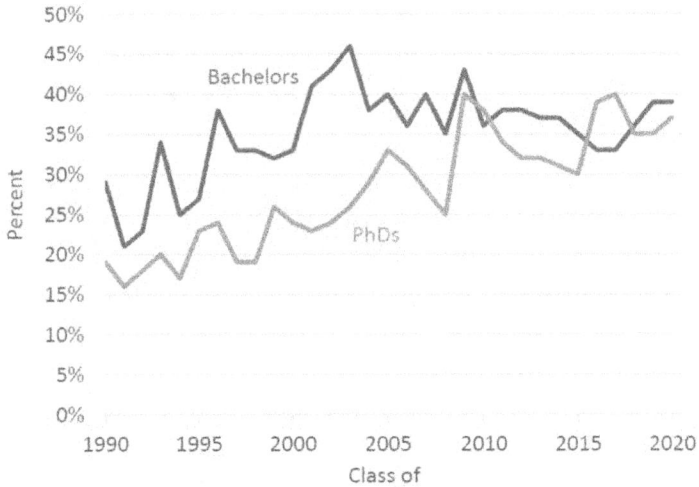

Figure 4.3. Percent of bachelor's degrees and doctorates in Astronomy earned by women, 1990–2020. Source: AIP Survey of Enrollments and Degrees.

More than one-fourth of the faculty members in astronomy are women (Table 4.1). Note that the 41% of assistant professors who are women (2018 and 2020) generally exceeds the proportion of PhDs earned by women. Thus, women are being hired above their relative availability. While only 19% of full professors are women, this is in line with the proportion of women earning PhDs in the past.

Table 4.1. Percent of Astronomy Faculty Members Who Are Women, 2003–2020

	Academic Year					
Academic Rank	2003	2006	2010	2014	2018	2020
Full professor	10%	11%	15%	15%	18%	19%
Associate Professor	23%	24%	22%	29%	26%	30%
Assistant Professor	23%	28%	30%	29%	41%	41%
Instructor/Adjunct	15%	15%	*	19%	30%	41%
Other Ranks	15%	21%	17%	22%	25%	27%
Overall	14%	17%	19%	19%	23%	27%

*There were too few individuals in this cell to report.
Table includes faculty members in departments that offer a degree in astronomy, but not in physics.
Source: AIP Academic Workforce Survey.

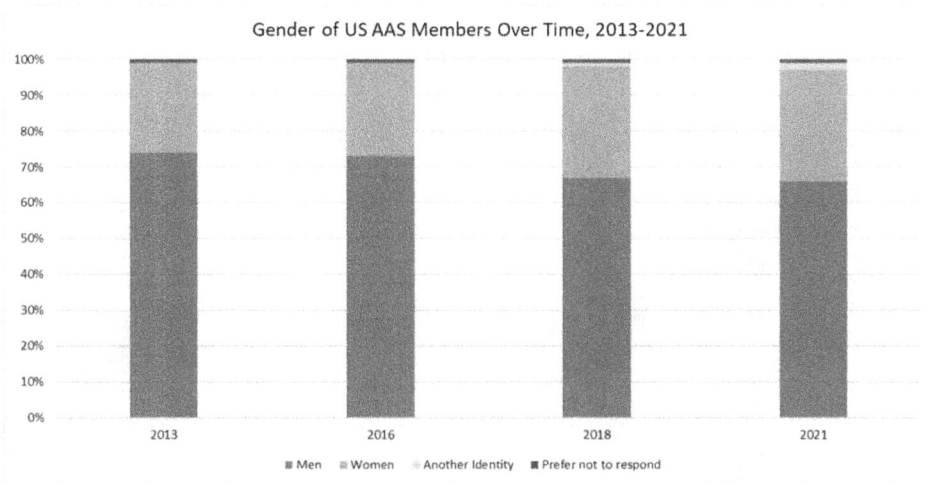

Figure 4.4. Gender of AAS members over time 2013–2021. Source: AAS Demographic Survey.

The percentage of women in the AAS membership is also growing (Figure 4.4). In 2013, one-fourth of AAS members in the US identified as women. In 2021, that percentage was 31%. Figure 4.4 also shows an increase in the proportion of AAS members who identify as a gender identity other than man or woman. Before 2018, there were not enough respondents who identified as another gender identity to report.[12] By 2021, 2% of responding AAS members in the US reported identifying as another gender identity. The AAS Demographics Committee survey also asks if respondents identify as transgender, but the number was not large enough to report without potentially violating confidentiality.

---

[12] In order to protect respondents' confidentiality, we do not report the number of people in a group if it is less than 5.

4.2.2 Race and Ethnicity[13]

People who identify as Hispanic/Latino, Black/African American, or Native American/ Alaska Native are underrepresented among all bachelor's degree recipients compared to their proportion of the US population who are 18–24 years old.[14] In 2019, the proportion of bachelor's degrees in all fields earned by people who identify as Hispanic/Latino (14.4%) or Black/African American (9.1%) exceeded that of people earning degrees in astronomy who identify as Hispanic/Latino (10.8%) or Black/African American (2.2%). American Indian or Alaska Native people earned 0.4% of all bachelor's degrees in 2019 and 0.4% of bachelor's degrees in astronomy (Figure 4.5). When translated to the number of graduates, fewer than 100 people who identify as Hispanic/Latino, Black/ African American, or Native American/Alaska Native earned bachelor's degrees in astronomy in 2019 (Figure 4.6). Fewer than 20 people who identify as Hispanic/Latino, Black/African American, or Native American/Alaska Native earned doctorates in astronomy in any single year from 2010 to 2020 (Figure 4.7).

A note about Asian/Asian American astronomers: In 2019, 7.5% of bachelor's degrees in astronomy/astrophysics were earned by people who identify as Asian.[15] In

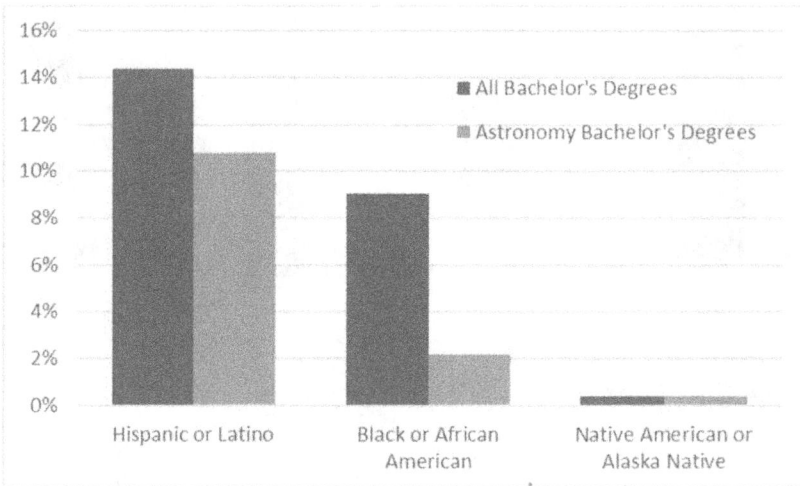

Figure 4.5. Proportion of bachelor's degrees earned by people who identify as Hispanic or Latino, Black or African American, or Native American or Alaska Native, 2019. Source: National Center for Education Statistics.

---

[13] The broad categories used in most surveys do not allow us to distinguish the different experiences of diverse people within these categories. Additionally, the racial and ethnic designations we use include only US citizens because definitions of race and ethnicity differ around the world.

[14] Among 18–24 year olds in 2020, 23% are Hispanic/Latino, 14% are Black/African Americans, 1% are Native American/Alaska Native, and 6% are Asian American. https://datacenter.kidscount.org/data/tables/11207-young-adult-population-ages-18-to-24-by-race-and-ethnicity#detailed/1/any/false/574,1729,37,871,870,573,869,36,868,867/68,69,67,12,70,66,71,7983/21595,21596

[15] IPEDS, the NCES data source, uses the term Asian rather than Asian American.

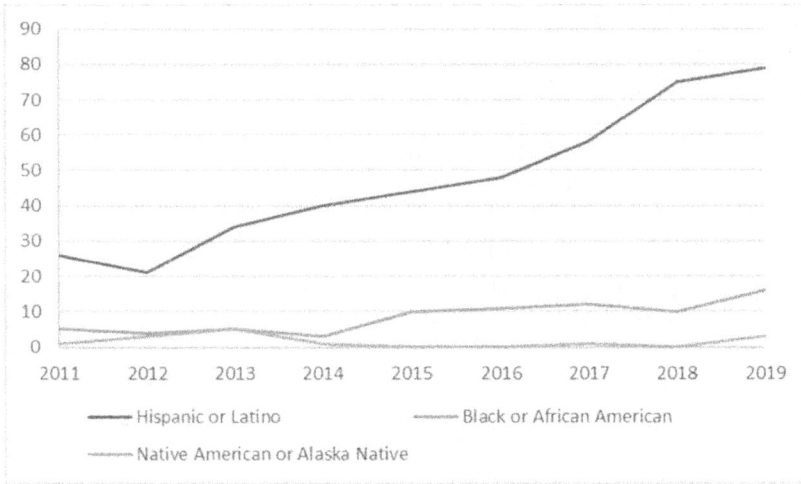

Figure 4.6. Number of Astronomy bachelor's degrees earned by people who identify as Hispanic or Latino, Black or African American, or Native American or Alaska Native. Source: National Center for Education Statistics.

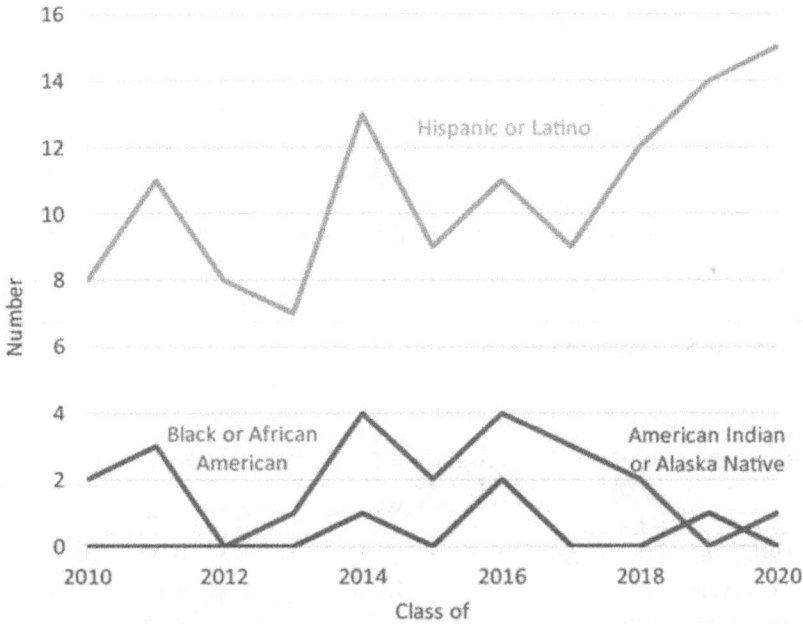

Figure 4.7. Number of Astronomy & Astrophysics doctorates earned by African American/Black, Hispanic/Latino, and Native American/Alaska Native People, 2010–2020. Source: National Center for Science and Engineering Statistics, NSF, Survey of Earned Doctorates.

2020, this percentage was 10.5% (NCES). Because the representation of Asian Americans in astronomy and astrophysics is higher than their representation in the overall population, Asian Americans are generally not considered to be under-represented. However, this does not mean that they are not marginalized, which would refer to the lack of equity and inclusion that they may experience. Unfortunately, data on inequality and lack of inclusion for Asian Americans in science are not widely available.

Over 90% of the US members of AAS in 2021 identified as White or Asian/Asian American (Table 4.2). As with degree recipients, Black/African American, Hispanic/Latino, and American Indian/Alaska Native astronomers are under-represented among US AAS members.

Table 4.3 shows the race and gender of astronomy and physics faculty members in 2021. (The number of astronomy faculty members in some of these groups is too low to report, so we have included physics faculty members.) For almost every race/ethnicity, the percentage of female faculty members is much lower than the percentage of male faculty members.

Table 4.2. Ethnicity of US AAS Members, 2021

Ethnicity	%	N
White	81	1369
Asian or Asian American	10	163
Hispanic or Latino	6	93
Black or African American	2	42
American Indian or Alaska Native	1	21
Native Hawaiian or other Pacific Islander	<1	3
Other	2	29
Prefer not to respond	4	66

Source: AAS Demographics Survey.

Table 4.3. Race and Gender of Physics and Astronomy Faculty Members, 2021

	Women	Men	Total
American Indian or Alaska Native	0.3%	0.3%	0.6%
Asian or Asian American	2.1%	6.3%	8.4%
Black or African American	0.8%	2.2%	3.0%
Hispanic or Latino	1.4%	3.2%	4.6%
White	17.9%	67.0%	84.9%
Other	0.6%	1.5%	2.1%
TOTAL	23.0%	80.4%	103.5%

Note: Total sums to more than 100% because people could select multiple races or ethnicities. Source: Source: AIP Faculty Member Survey

Table 4.4. Disabilities Among US AAS Members, 2021

Disability	%	N
I have a mental illness	7	120
I have an autoimmune or pain disorder, or other chronic condition	6	107
I am deaf or hard-of-hearing	4	73
I have disabling allergies, asthma, or other environment sensitivities	2	27
I am neuroatypical	4	68
I have difficulty seeing even when wearing glasses	1	22
I have serious difficulty standing, walking, or climbing stairs	1	23
I have a cognitive or learning disability	1	24
Other disability	3	44
None of the above	74	1129
Prefer not to respond	5	84

Source: AAS Demographic Survey.

Table 4.5. Sexual Orientation of US AAS Members, 2021

Orientation	%	N
Heterosexual or straight	83	1385
Gay or lesbian	3	57
Bisexual	5	91
Other	3	51
Prefer not to respond	6	94

Source: AAS Demographic Survey.

4.2.3 Disability

It is important to remember that many astronomers with disabilities make significant contributions to the field. As seen in Table 4.4, the most commonly reported disabilities among AAS members were mental illness, autoimmune or pain disorders, or other chronic conditions. One in twenty-five AAS members reported being deaf or hard-of-hearing. Almost three-fourths of the membership reported no disability (Table 4.4).

4.2.4 Sexual Orientation

More than 10% of the US AAS members reported their sexual orientation as gay or lesbian, bisexual, or some other orientation (Table 4.5).

According to a 2022 survey conducted by Gallup, seven percent of US adults identify as LGBT+.[16] Among astronomers, the percentage who identify as LGB+[17]

---

[16] https://news.gallup.com/poll/389792/lgbt-identification-ticks-up.aspx
[17] As noted above, the AAS/AIP survey asked about transgender identity on a separate question from sexual orientation, but we cannot report their numbers without violating confidentiality.

is 11%, which is not an indicator of under-representation. However, just knowing the representation of a group does not tell us whether that group is marginalized, or experiences inequity and exclusion in astronomy and astrophysics. In the next section, we will examine data from surveys that do look for evidence of marginalization for most of the groups we have discussed so far.[18]

4.3 Equity and Inclusion in Astronomy

Diversity can be measured by examining the representation of groups in a scientific field. Equity and inclusion cannot be measured in the same way as representation. Even if a group is represented in proportion to some external standard (e.g., the same percentage in astronomy as in the general population), members of that group could still encounter inequity and exclusion in astronomy.

In order to test whether members of groups encounter inequity and exclusion in astronomy, we use measures of experiences such as staying in the field after graduate school or experiencing sexual harassment. We can look to see whether being a part of a particular group increases or decreases the likelihood of having that particular outcome or experience.

We use data from the Longitudinal Study of Astronomy Graduate Students (LSAGS) to test measures of inequity and exclusion in astronomy. This study arose from the 2003 Women in Astronomy Conference in Pasadena, where it was noted that a majority of young members of the AAS were women. The astronomy community wishes to make every effort to retain young women in astronomy, so AAS worked jointly with AIP to conduct a longitudinal study that would pinpoint the factors that contribute to retention in general, with a focus on differences between women and men. The LSAGS follows a cohort of people who were graduate students in astronomy or astrophysics during 2006–2007. The first survey was conducted during 2007–2008, the second during 2012–2013 and the third during 2015. The analysis presented in this paper used a subset of 300 respondents, all of whom had PhDs in astronomy, astrophysics, or a related field at the time of the third survey.

Below, we examine experiences that contribute to retention in astronomy and physics to measure inequity. We will answer the question: are the experiences that tend to contribute to retention equitably distributed among different types of people? We do this with statistical models that control for many variables so that we can isolate the effects of one variable at a time.

After examining retention, we will use experiencing harassment and discrimination as a measure of exclusion. We also use statistical models to determine which types of people are the most likely to report harassment. Harassment and discrimination are types of exclusionary behaviors that can be used to send the message that the recipient does not belong in astronomy.

---

[18] Unfortunately, we do not yet have survey data that allows us to document the effects of having a disability on inclusion and equity in astronomy.

4.3.1 Retention in Astronomy

In our statistical models explaining attrition, we tested effects of the imposter syndrome,[19] mentoring and advising during graduate school, the so-called "two-body problem" that occurs when a couple needs to find two jobs in the same geographic area, and the gender of the respondent (Ivie et al. 2016). For this analysis, we added race/ethnicity and sexual orientation to our previous models.[20]

As with our previous research, we ran a logistic regression examining employment in or out of astronomy at the time of the second survey. Our initial model included the following as independent variables:

- The respondent's imposter syndrome rating[21]
- The number of years since the respondent earned their PhD
- Whether the respondent had ever taken a postdoc
- Whether the respondent had relocated for a partner
- Whether the respondent had limited their career for a partner
- Whether the respondent had a mentor in graduate school
- Whether the respondent changed advisors during graduate school
- The respondent's rating of their advisor
- Whether the respondent identified as Asian or Asian American
- Whether the respondent identified as Black, Hispanic, Indigenous, or "other" ethnicity
- The respondent's gender
- Whether the respondent was gay, bisexual, or another non-heterosexual sexuality
- Whether the respondent had been harassed at school or work

We did a backward stepwise regression eliminating the variable with the largest *p*-value at each step. We found three variables with direct effects on attrition from the field of astronomy (Figure 4.8).

- Respondents who had *relocated for a spouse or partner* were **2.4 times** *less* likely to be employed in astronomy at the time of the second survey.

---

[19] Our measure was based on the original concept of the imposter syndrome. See Clance, P. R., & Imes, S., 1978, Psychotherapy Theory, Research and Practice, 15, and Clance, P. R., 1985 *APA PsycTests*, https://psycnet.apa.org/doi/10.1037/t11274-000. The imposter syndrome concept has been critiqued and other names for it, such as the imposter phenomenon, have been suggested. See https://hbr.org/2021/02/stop-telling-women-they-have-imposter-syndrome for a critique relating this "phenomenon" to bias and exclusion.

[20] In the second survey (conducted in 2012–2013), we asked about race/ethnicity and sexual orientation. For race/ethnicity, respondents were asked "What is your race or ethnicity?" and could check multiple options from American Indian or Alaska Native, Asian or Asian American, Black or African American, Hispanic or Latino, Native Hawaiian or other Pacific Islander, White, and other. For sexual orientation, we asked, "Do you consider yourself to be" with response options of heterosexual or straight, gay or lesbian, bisexual, and other. Respondents could choose only one option. Information on respondents' gender was collected in 2007. At that time, only two gender options were offered, and we do not have a measure of whether respondents identify as transgender.

[21] See https://www.aip.org/statistics/reports/women-and-imposter-syndrome-astronomy for a description of the items used in this rating.

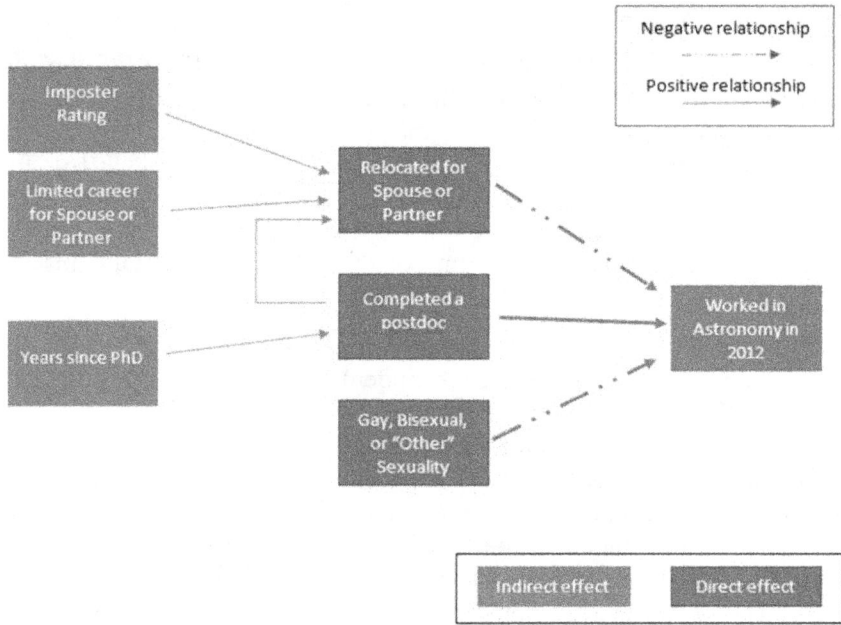

Figure 4.8. Direct and indirect effects on persistence in Astronomy, Longitudinal Study of Astronomy Graduate Students, Second Survey.

Table 4.6. Selected Characteristics of LSAGS Respondents

Characteristic	% Identifying with
Sexual orientation: Gay, Lesbian, Bi, or "Other"	4%
Asian or Asian Americans	14%
Race/Ethnicity: Black, Hispanic, Indigenous, or "Other"	6%
Ever personally harassed or heard about or witnessed harassment	33%
Personally experienced harassment or discrimination	30%
Woman	40%

Source: Longitudinal Study of Astronomy Graduate Students, AIP

- Respondents who had *completed a postdoc* were **2.9 times *MORE*** likely to be employed in astronomy at the time of the second survey.
- Respondents who *identified as gay, bisexual, or some other sexuality* were **11.3 times *less*** likely to be employed in astronomy at the time of the second survey.

The direct effect of sexual orientation illustrates inequity in astronomy. Astronomers who are lesbian, gay, bisexual or another non-heterosexual identity (LGB+) were more likely to leave the field than other astronomers. The reasons for this are not discernable from the results of the survey. We note also that although a higher percentage of AAS members are LGB+ as in the US population, there is a lower percentage of LGB+ astronomers among the LSAGS respondents (Table 4.6).

Further research is needed in two areas: first to understand whether the experiences of LGB+ astronomers in this cohort are unique, and second, to understand the lived experiences of LGBT+ astronomers, and how these experiences contribute to persistence or attrition in the field.

There were also indirect effects of some of the variables on attrition from the field. We found that the imposter syndrome rating, the years since earning one's PhD, whether a respondent had limited their career for a spouse or partner, and completing a postdoc all had indirect effects on working in astronomy or physics. "Indirect" means that a variable has an effect on the likelihood of experiencing or being in a category that directly affects the dependent variable. For example, people with a high imposter score were more likely to relocate for a spouse or partner, and therefore were less likely to be working in the field.

We tested similar models on the data from the third survey, which occurred eight years after the first survey. We used the third survey to try to ascertain more about specific actions people's advisors took that may have contributed to persistence or attrition. However, none of the variables we tested had statistically significant effects. Instead, we found that working in the field in 2012 and having completed a postdoc were the best predictors of working in astronomy/physics in 2015.

4.3.2 Encountering Harassment and Discrimination

In the second LSAGS survey, we asked respondents if they had ever encountered harassment or discrimination at school or work.[22] Using statistical models, we examined the impact of race, gender, and postdoc status on encountering harassment or discrimination, which are types of exclusionary behaviors. We included intersectional data for race and gender. For our analysis, we classified respondents into three racial groups: white (those who selected white only), Asian (those who selected Asian only or Asian and white), and Black, Hispanic, Indigenous or other minoritized persons (those who selected Black or African American, Hispanic or Latino, American Indian/Alaska Native, Native Hawaiian/Pacific Islander, or "other.").

Black/African American, Hispanic/Latino, Indigenous (American Indian, Alaska Native, Pacific Islander, Native Hawaiian) and other minoritized people (abbreviated here as BHIO), along with women, were much more likely to report having encountered discrimination or harassment than white men. Being a postdoc at the time of the survey increased the likelihood of reporting harassment or discrimination for women and people who identify as BHIO (Figure 4.9). However, there was no statistically significant difference between Asian/Asian American astronomers and white astronomers in reporting harassment and discrimination. Thus, we see that intersectionality between gender race/ethnicity increases the likelihood of experiencing discrimination or harassment. These results support the conclusion that white women and BHIO people of all genders face exclusionary behaviors in astronomy.

---

[22] For a description of the types of behaviors respondents described as harassment or discrimination, see https://www.aip.org/statistics/reports/exploring-harassment-and-discrimination-experiences-astronomy

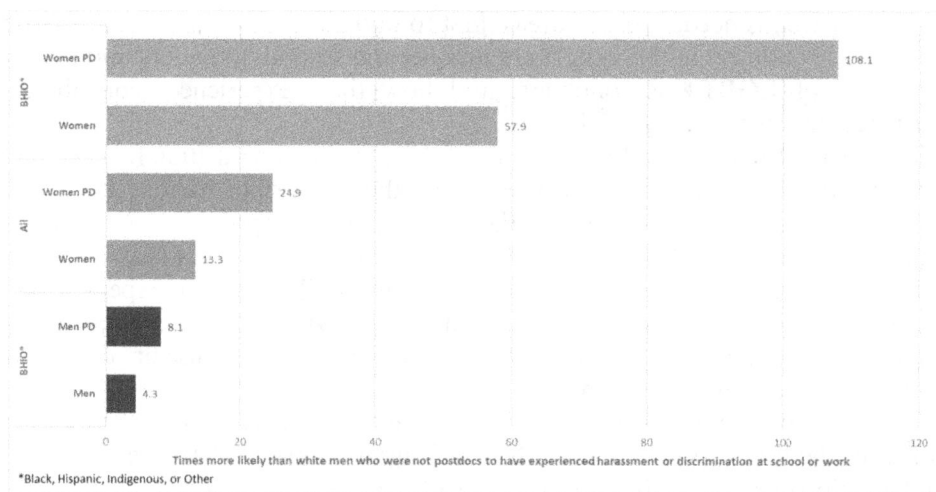

Figure 4.9. Experiencing harassment and discrimination by gender and race/ethnicity, Longitudinal Study of Astronomy Graduate Students, Second Survey.

These types of exclusionary behaviors are an assertion of power and send the message that the receiver is not welcome in astronomy/physics.

4.4 Discussion and Conclusion

We have presented evidence that specific groups of people (Black/African American, Hispanic/Latino, American Indian, Alaska Native, Pacific Islander, and Native Hawaiian) are not only numerically under-represented, contributing to a lack of diversity in astronomy, but both men and women in these groups experience exclusionary behaviors in the form of harassment and discrimination. Furthermore, we have seen that women in general are under-represented and experience harassment and discrimination. Awareness of the effects of exclusionary behaviors such as these has increased since the publication of the National Academies report on sexual harassment in 2018.[23] Clearly more needs to be done to eliminate these behaviors, thereby reducing the negative effects they have on people in science.

Finally, we have seen that LGB+ astronomers in the Longitudinal Study of Astronomy Graduate Students (LSAGS) were more likely to leave the field. The data in this study do not provide any additional information about why LGB+ astronomers in this cohort were more likely to leave the field. In addition, none of our data sources provide information on the types of inequity and exclusion that astronomers with disabilities may face. These are areas in which additional research is necessary. We would like to note that surveys may not be the best way to collect

---

[23] National Academies of Sciences, Engineering, and Medicine. 2018, *Sexual Harassment of Women: Climate, Culture, and Consequences in Academic Sciences, Engineering, and Medicine* (Washington, DC: The National Academies Press), https://doi.org/10.17226/24994.

data on people's lived experiences of inequity and exclusion. Surveys are not a good way to examine different experiences between people in small groups, such as the differences among the subcategories of people who are Hispanic/Latino. Instead, these are areas that are best explored with qualitative research using interviews or focus groups.

The data in this paper focus only on limited measures of inequity and exclusion. There are many others still to be investigated. Some of these may include pay gaps, reduced access to opportunities, and inequitably distributed resources to conduct research. For example, our analysis found no differences between white and Asian/Asian American astronomers in reporting harassment and discrimination. There are still many areas in which Asian/Asian American astronomers—and astronomers in other groups—may experience inequity and exclusion.

It is necessary for astronomy to continue to collect data on inequity and exclusion. Without these types of data to illuminate problem areas, efforts to increase diversity are unlikely to succeed. In order to recruit and retain the diverse astronomers who are needed to advance the field, we must know where the problem areas lie so that they can be addressed. As long as astronomy lacks equity and inclusion, some people will be dissuaded from pursuing it. An equitable and inclusive environment should lead to fewer instances of discrimination, higher rates of retention, and eventually more diversity.

Acknowledgements

The authors would like to thank our colleagues in the SRC who contributed data and analysis to this chapter: Patrick Mulvey, Starr Nicholson, Jack Pold, Anne Marie Porter, and John Tyler. We also thank the National Science Foundation for its support of the Longitudinal Study of Astronomy Graduate Students (Grant 1347723).

References

Campbell, L., Mehtani, S., Dozier, M., & Rinehart, J. 2013, PLoSO, 8, e79147
Freeman, R. G., & Huang, W. 2014, Natur, 513, 305
Hong, S., & Page, S. E. 2004, PNAS, 101, 16385
Ivie, R., White, S., & Chu, R. Y. 2016, PRPER, 12, 020109
Puritty, C., Strickland, L. R., Alia, E., et al. 2017, Sci, 357, 1101
Rock, D., & Grant, H. 2016, HarBRv, https://hbr.org/2016/11/why-diverse-teams-are-smarter

Part II

Actions in the Astronomy and Astrophysics Community

Introduction to Part II: Actions in the Astronomy and Astrophysics Community

This section showcases community-led actions in advancing professional inclusion. In recent years, there have been an increasing number of efforts within the astronomy and astrophysics community to address and promote diversity, equity, inclusion and access (DEIA). Many of these efforts primarily focus on remedying concerns for students, or in academic education. These actions are often targeted within a single university or even more narrowly, a single department. A number of these kinds of activities have been written about in the Bulletin of the American Astronomical Societies' journal edition dedicated to white papers on the state of the profession written for the 2020 Astronomy and Astrophysics Decadal Survey (For a link to this collection, see footnote 11 in chapter 15 of this ebook).

However, here we have chosen to highlight examples of DEIA actions that are focused on innovative activities being undertaken by *scientific collaborations*. We focus on these structures because, as projects become larger, more complex and more expensive, large collaborations are becoming an increasingly important part of scientific culture. As significant organizing structures in the field, scientific collaborations represent places where cross-fertilization of ideas, policies and methods can lead to significant cultural changes. In addition, collaborations provide consequential opportunities for leveraging a broader base of researchers with common interests and goals for the advancement of students and early career scientists. Having these critical structures function in an ethical and principled way with respect to equity in science can promote broad cultural shifts throughout the field. Furthermore, in some cases, collaborations, which cut across different universities (and indeed different countries) may be in a position to achieve more than what can be accomplished within a single institution or department due to barriers like gatekeeping or administratively perceived risks of embracing diversity goals.

Although members of these collaborations are individual scientists, affiliated with distinct universities or other institutions, they are working toward a (or a set of) particular scientific goal(s) and share a mission that sees advancing DEIA principles as adding meaningful value to their work. These collaborations have identified ways to build partnerships across their individual affiliations that promote a common structural culture where engagement with DEIA principles has been normalized as how the collaboration operates day to day. In the article *"Overview of SDSS IV's COINS"*, the authors point out the importance of data collection as a way of understanding the working climate within the collaboration. The articles, *"Leveling the Playing Field"* and *"Anti-Black Racism Workshop"* represent initiatives where collaborations were able to leverage philanthropic funding sources to promote opportunities to engage DEIA activities that supported not only their own collaboration, but other adjacent collaborations. Each set of authors recognizes and presents both achievements and lessons learned from their significant efforts that can be an important source of information for the reader.

An Astronomical Inclusion Revolution
Advancing diversity, equity, and inclusion in professional astronomy and astrophysics
Dara Norman, Tim Sacco and Dorian Russell

Chapter 5

Overview of SDSS IV's COINS: Achievements and Shortfalls

Amy M Jones, Rachael L Beaton and The Committee on INclusiveness in SDSS (COINS)

5.1 Introduction

The Sloan Digital Sky Survey (SDSS) is a multi-phase international collaboration that has been in operation since 1998 (York et al. 2000). With the goal of mapping the Universe at all scales, SDSS has collected, analyzed, and released data from several instruments and campaigns. It has been divided into phases, with each phase having different dedicated surveys and management teams. The fourth phase, SDSS-IV, started collecting data in 2014 and its last data release (DR17) was in 2021 (Blanton et al. 2017). SDSS-IV had three core programs: APOGEE-2, eBOSS, and MaNGA (for more details, see Blanton et al. 2017). SDSS-V began in 2020 and is currently on-going (Kollmeier et al. 2017). SDSS has over 1000 member scientists and staff from five continents and is actively taking data in the United States and Chile. In Figure 5.1 we show the geographic distribution of SDSS-IV members from our demographic surveys. Thus, SDSS brings together many distinct social and scientific cultures and backgrounds as part of one large collaboration.

The Committee on INclusiveness in SDSS (COINS) was established as a standing committee in the fourth phase of SDSS (SDSS-IV). COINS formed in 2016 by merging two other committees, the Committee on the Participation of Women in SDSS (CPWS) and the Committee on the Participation of Minorities in SDSS (CPMS), with the mandates from both. CPWS was initiated in 2013 after a disparity in the gender balance of the SDSS leadership structure was identified by the Sloan Foundation (see discussions in Blanton et al. 2017, 2019). CPWS conducted the first demographic survey of SDSS-IV in 2014 and we have regular surveys to monitor the make up of the collaboration and project over time. Results from the surveys are used to inform our recommendations to SDSS management and influence COINS' policies. CPMS was formed to address the under-representation of minorities in the survey and to help recruit and retain underrepresented

Geographic Location

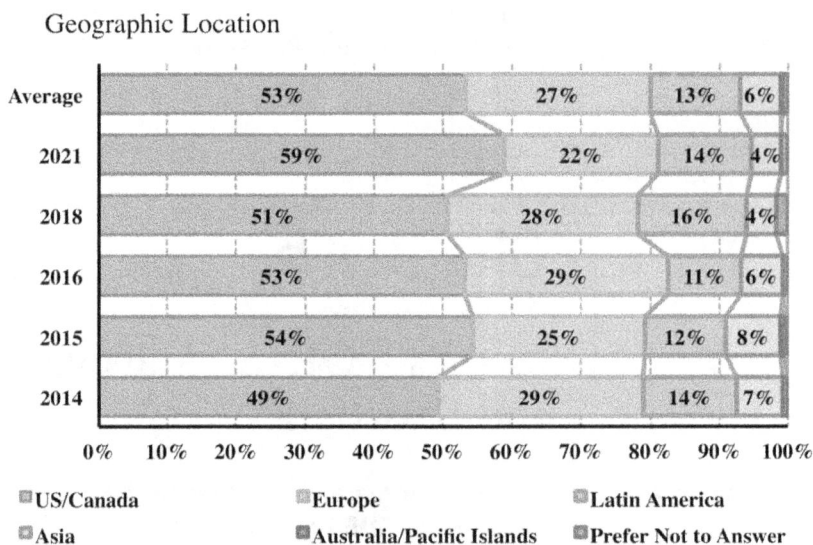

Figure 5.1. Distribution of responses from the SDSS-IV demographics surveys on geographic location. The average over all the SDSS-IV surveys is given in the top row, with the results from each individual survey given in the subsequent rows (from top to bottom: 2021, 2018, 2016, 2015, and 2014). Roughly half of respondents live or work in the US/Canada, about a quarter in Europe, one seventh in Latin America, and less than one tenth in Asia and Australia/Pacific Islands.

minority talent in SDSS (Blanton et al. 2017, 2019). Having these committees formed by SDSS-IV leadership gave them more legitimacy and power within SDSS to enact change. In addition to the support from the Sloan Foundation, COINS has continually had support from the management committee and SDSS leadership in general. It is important to have both the support from the funding agency and leadership alongside the efforts coming from the members, themselves, that address issues from the ground up.

This contribution reflects on the successes and shortcomings of COINS in SDSS-IV (2013–2021), focusing on three key areas:

- Demographic Surveys
- Gender Balance in Leadership
- Role in Collaboration Meetings

We conclude with a broader span of achievements and limiting factors in making SDSS a more inclusive, accessible, and welcoming collaboration.

5.2 Demographic Surveys

One of the key roles of COINS is to administer and analyze regular demographic surveys of the collaboration. So far we have had surveys in 2014, 2015, 2016, 2018, and 2021, spanning the lifetime of SDSS-IV (Lundgren et al. 2015; Lucatello & Diamond-Stanic 2017; Jones et al. 2023).[1] On average, there are about 250

---

[1] All five surveys can be found at https://github.com/sdss/coins/tree/main/materials#demographic-surveys

respondents to each survey. The set of questions from 2014 has been maintained to study trends with time, though more questions have been added to later surveys. Data from the demographic survey, that includes open-answer responses from the SDSS members, is crucial to assessing the climate and identifying factors for improvements. For more details on how the surveys were conducted, please see Lundgren et al. (2015), Lucatello & Diamond-Stanic (2017); Jones et al. (in preparation).

To highlight the importance of getting feedback from the full SDSS community and not a smaller subset like leadership, we can look at the responses to the question about how much does one agree with "SDSS fosters an inclusive environment." Leaders were more likely to agree with this statement than respondents that did not consider themselves to be in a leadership role (see Figure 5.2). This illustrates the need for regular community wide surveys to understand the climate of the collaboration to hopefully be less biased and more representative of all SDSS members.

One of the concerns with designing and analyzing the surveys is that COINS is composed of scientists and technical staff without any formal training in social sciences. For the 2014 survey paper, then CPWS did consult with a social scientist in the American Institute for Physics (AIP). In general, it can be difficult to interpret some of the findings since this type of analysis is outside our expertise. COINS members have had to develop these skills on our own initiative.

Another issue has been the lack of resources to take the results from the survey(s), analyze them in multiple ways on various axes, and publish our findings. With more time to spend on the survey results and additional expertise, COINS could take better advantage of the rich results from the surveys. The data could be sliced on multiple demographic axes and monitored over time.

The survey has also expanded over time. The more recent survey(s) included a section on comfort levels when asking or answering questions in different

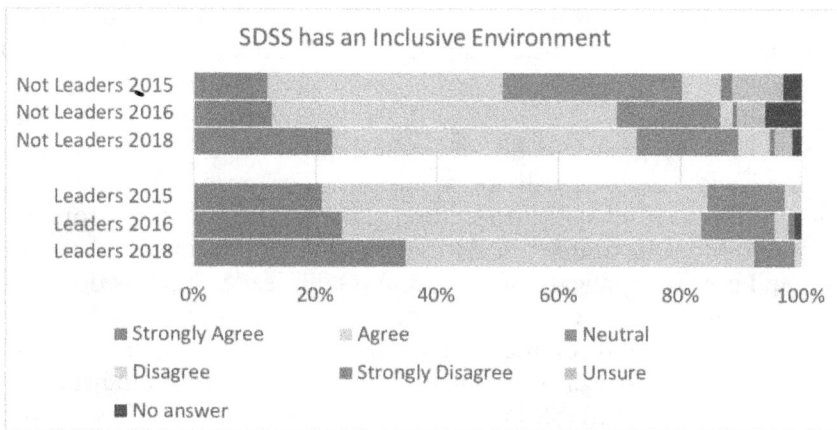

Figure 5.2. Distribution of responses from the SDSS-IV 2015, 2016, and 2018 demographics surveys about if they agree with the statement that SDSS is an inclusive environment. Respondents are split into leaders and not leaders, based on if they had self-identified as being in a leadership role in SDSS.

collaborative settings, additional demographic questions, and a section on the impact of the COVID-19 pandemic. These are among the many possibly interesting facets of the surveys that have not yet been fully explored due to the lack of resources. After completing some analysis, progressing to publication quality also takes time and effort, in particular in finding comparative data sets and being cautious about our interpretations. Additionally, in order to keep the raw results confidential, only those members who were in COINS at the time of the survey are allowed to access the raw response data. This makes it increasingly difficult to analyze and publish results as members rotate off and access to the raw data becomes more limited. Once the response data is aggregated, it is accessible to all the COINS members. Access to the raw data is required, however, to aggregate responses on a new demographic axis or set of axes that had not been considered previously.

A major limitation of COINS, especially in regards to analyzing demographic surveys, is that service oriented tasks are often considered lower priority than other responsibilities. All academic positions have to balance research, functional or teaching duties, and both institutional and community-wide service. COINS is a service group that functions for a collaboration and falls into a grey area in terms of how or if institutions value the work. Only having a handful of people working on analyzing and publishing results from the surveys for a small fraction of time leads to slow progress; for example, the current demographic paper has taken over five years and had a major re-scope of the content two years ago. Service work needs to be treated just as equal as research in terms of funding and career progression.

5.3 Gender Balance in Leadership

One of the first areas COINS (known as CPWS at the time) was asked to address was the lack of women in SDSS leadership. Thus, the first demographic survey established a base set of questions related to leadership (see Lundgren et al. 2015). Jones et al. (in preparation) compares the 2014, 2015 and 2016 data to see if there are any trends in these demographics; future efforts will extend this analysis to include the 2018 and 2021 surveys.

Two recommendations were made to the management committee at the beginning of SDSS-IV. First was to fill all leadership positions through a formalized approach and with an open call for applicants. The open call should include a description of the position and expectations and be sent to an appropriate email distribution list(s), with ample time for recipients to consider the position, ask questions, and prepare applications. This process fills leadership positions in a more transparent manner and makes access to these roles open to the full SDSS collaboration. After this recommendation, the majority of the unfilled leadership positions were done through an open call. However, at this time, many of the management positions in SDSS-IV were already filled. SDSS-V begun with these practices in place, so there should be more of an impact in the SDSS-V leadership demographics. From the demographic surveys, we have seen that this is only one of many paths to leadership roles and it only included about 20% of all responses

(Jones et al. in preparation). In the demographic surveys, we purposefully allowed respondents to self identify as being in a leadership role. We defined leadership as "any role whose tasks or responsibilities include making decisions that affect other people and the survey, organizing regular project discussions or meetings, professional mentoring, or influencing/directing others in their tasks". This would include all forms of leadership in scientific collaborations, not just the formal management. From the survey results, many of the leadership roles were defined around work already being done and did not go through a formal application process.

Leaders chosen through an open call were mostly unbiased by gender. However, relatively more men become leaders based off of work they were already doing. Thus, the overall gender balance of leaders remains biased towards men. COINS still highly recommends following the policy to have as many positions filled through an open call with a formal, transparent process. Acknowledging that other paths into leadership will continue to exist, we need to understand how to address and reduce any possible gender bias.

The second recommendation was to have a deputy position created and filled for all of the top leadership positions. The deputy would gain important leadership skills and would be more ready to take on a higher leadership position at a later date. Many deputy positions were advertised through an open call. A major challenge was that many of these were never filled, having received no applicants. Through the demographic surveys, COINS has data on why SDSS members are reluctant to apply for leadership positions. Because most roles come with no additional support (like salary or travel support), leadership positions are seen as extra work and responsibility with little to no added benefits. We note that leadership roles are not always perceived to have inherent value or clear connections to career advancement. This is especially challenging because a plurality of SDSS participants, as measured on the demographic survey, are on term-limited contracts or soft-money, where the work required for a leadership role may be at conflict with work plans for their grants. While most positions still advertise for a deputy, until these are better incentivized for the SDSS community, there will probably be only a few deputy positions filled. COINS would like to further explore how leadership roles in SDSS have impacted careers.

5.4 Role in Collaboration Meetings

SDSS holds an annual collaboration-wide conference. These collaboration meetings are typically in person at locations that rotate between continents; due to the COVID-19 pandemic, however, the meetings in 2020 and 2021 were fully virtual. Most of the communication and interaction among SDSS members is done remotely due to SDSS not being localized but world-wide (Figure 5.1), and the collaboration meetings are unique opportunities for the collaboration to come together in person (pre-pandemic) or virtually (pandemic) to meet each other and form a sense of community. COINS has played a significant role in helping organize and run the meetings and its associated activities. We have a critical focus on accessibility and inclusion for all, in particular making sure new or more junior SDSS members feel welcomed into the collaboration community in an intentional way.

For the accessibility of the conference, COINS has drafted guidelines[2] for what should be considered with regards to the venue (when in person), organizers, and attendees. Additionally a member of COINS is typically a member of the organizing committee to ensure that such practices are followed. A simple solution to ensure accessibility needs are met is to have a place on the website and registration form where attendees can express any accessibility concerns and then these can be addressed directly. One limitation is that many of the decisions that would affect accessibility are made without COINS involvement, or are not feasible depending on the location or venue. This demonstrates why inclusivity and accessibility should be folded into decision-making at the highest level. One example occurred at a buffet-style conference dinner, where dietary restrictions were not considered. In this case there was only a single vegetarian dish and, because it was not reserved for vegetarians, it was often empty. The vegetarian and vegan participants had little to no options. If on the registration form there had been a place to state any dietary restrictions and this information had been passed on to the caterers, then the situation could have been avoided. As organizers it can be difficult to think of all possible scenarios, such as making sure everyone is included and has options at a conference dinner. Hence, it is important to have a person or a group dedicated to accessibility concerns that is involved from the beginning.

For the last three meetings, COINS developed a guide to inclusive chairing[3] and hosted training sessions with session chairs. During a typical session with multiple presentations, the chairs have a great influence over how inclusive the session feels to both the presenters and the attendees; especially so in the virtual setting. Because this role is more vital and more demanding in a virtual environment, COINS has recommended having two chairs per session; roles are then divided such that one helps introduce the speakers and keeps track of the timing and the other focuses on questions that may be posed through multiple channels (e.g., live, on Slack, on Zoom). In the virtual meetings, the chairs in every session are also responsible for posting the reminder to attendees to update their display name to act as a virtual name badge, this includes stating one's pronouns and other SDSS roles. We have gotten some feedback that these rules and reminders can be annoying to some, but we have also received a great deal more positive feedback in regards to how this has helped marginalized groups feel more welcome and how it expresses respect for speakers and participants.

In the virtual conferences, we also take advantage of the chat feature during the meeting and asynchronous communication on Slack to collect questions. We strongly discourage using the chat for conversations during the presentations, which can be distracting for both the presenter and audience. After the talk, questions can be posed in chat or on other platforms like Slack. Every talk had its own thread in Slack where participants can post questions to the speaker; speakers are encouraged to answer these questions in Slack later. This can help those that cannot attend the

---

[2] https://github.com/sdss/coins/blob/main/documents/md/meeting_accessibility.md
[3] https://github.com/sdss/coins/blob/main/documents/md/chairing_guidelines.md and https://github.com/sdss/coins/blob/main/documents/md/virtual_conferences.md

meeting live to still engage with the meeting. As SDSS moves towards hybrid meetings, such asynchronous communication would be helpful for those that cannot attend. COINS emphasizes a few key elements for chairing question and answering sessions: (1) pausing to allow participants to collect their thoughts, (2) encouraging questions from more junior attendees, and (3) asking participants to introduce themselves before asking their question. These recommendations have been normalized such that at almost every session, the chair will remind the attendees that they are pausing and hoping to have some questions from early career researchers.

During the COINS training on inclusive chairing practices, sometimes there was resistance from chairs. Instead of recognizing COINS' efforts to enforce inclusive practices, some chairs wanted to run the sessions their own way. Feedback after those "self-managed" sessions tend to show that speakers and participants prefer our guidelines so they can better focus on and engage with the content.

COINS also runs several activities at the collaboration meeting aimed at engaging early career researchers and new or junior members. At in person meetings, COINS held a "welcome breakfast" where more junior members would sit next to more senior members of the collaboration. This became difficult to schedule and changed into a meet and greet session either during the opening reception or on the first day during a longer break. The current version we call "speed geeking", where junior members are introduced to senior members (and vice versa) for 3 min before shuffling. This was a way for both groups to quickly meet many collaboration members in a fun, albeit sometimes chaotic way. In the virtual meetings, we changed the format slightly to use break out rooms in Zoom with groups of 3–5 people that would be re-organized every few minutes. The senior members and management were supportive of this event, which is crucial to it being successful and in making management more accessible.

In addition to speed geeking, COINS often ran a BINGO activity to encourage more interactions between members. Participants received a BINGO card that contained experiences common to SDSS (serving on a committee, visiting the observatory, plugging a plate, among others). The primary goal was to help newer members approach more senior members, but anyone was welcome to participate. A secondary goal was to articulate the many roles required for the collaboration to be successful. Once someone acquired a signature in each square, they were eligible for a prize. This activity was hard to port into a virtual environment, and so has only occurred during the in person meetings. Because COINS does not have its own budget, we also had to rely on getting SDSS items that already existed or could be crafted, usually from Education and Public Outreach. One year we also tried having a buddy system for new members, but this became difficult to manage and we did not have enough volunteers to meet the needs of new members.

COINS also facilitated townhalls, career panels, and inclusivity trainings during breaks between scientific sessions that were open to all members. The townhalls were sometimes topical, e.g., on the Code of Conduct, but more often were used as a space for discussion between SDSS members and COINS. These discussions were often fruitful and helped us connect with various groups within SDSS. Again management was supportive and usually attended these townhalls. The career panels were aimed towards early career researchers and were a place to advertise several different possible career paths undertaken by current or former SDSS members. The trainings, which have so far only happened at the in person meetings, covered topics like inclusive leadership and bystander training. A COINS member had led similar trainings before joining SDSS and we were able to use their expertise to provide them at the SDSS meetings.

For all the COINS events at conferences, we have learned that it is critical to have good communication with the meeting organizers from the beginning and to have support from management. Sometimes, COINS has been asked to contribute to planning much too late in the process, when many of the organizing decisions had already been made, such as the venue, outline of the meeting schedule, plenary speakers, etc. COINS then has to spend even more energy and time trying to adapt these plans (if possible) to make the events as inclusive and accessible as is feasible. Often there was little room in the schedule for the various activities and events mentioned earlier; we have received complaints about it being difficult to attend all events members are interested in due to timing. Many of our activities are time critical, for example the chairing guidelines meeting needs to happen before the conference starts, and "speed geeking" should take place during the first day to help new members feel welcome and part of the SDSS community. Additionally COINS needs to be given the power to enforce the practices we suggest. COINS has put in a great deal of effort and research into our recommendations; it can be frustrating and defeating when others choose to ignore our guidelines because they think they are invalid.

In order to gauge how successful the meetings were in these aspects and others, COINS and the meeting organizers administer post-meeting surveys. Overall we have received positive feedback about the COINS activities and how the meetings were run. Also from the SDSS demographic surveys, we see a trend where more members have found SDSS to be an inclusive environment from 2015 to 2018, in the time-frame where COINS have been actively involved in meetings (Figure 5.2).

5.5 Conclusions

In the previous sections we discussed in detail three areas COINS has focused on in SDSS-IV. We conclude with brief descriptions of achievements and shortfalls as COINS to help SDSS be more inclusive, accessible and welcoming.

In SDSS-IV, COINS has had several successes, some of which are mentioned in more detail earlier.

- Inclusion efforts recognized as an official role in SDSS:
  - Time spent in COINS is now recognized as fundamental to the infrastructure of the survey and can count towards so called "Architect" status, which grants extra publication rights.
  - COINS web documentation is formally part of data release efforts formally and its documents are considered a product of SDSS.
  - COINS presents a plenary talk at each collaboration meeting.
  - There is a section on COINS in the annual reports to the Sloan Foundation.

- COINS continued into SDSS-V as a standing committee. The chair is now officially recognized as a member of the Management Committee.
- Material and documentation for SDSS and the astronomy community:
  - COINS was consulted on the Code of Conduct[4] that was adopted in 2018 and drafted supporting documents.[5] We also promoted the Code of Conduct with a dedicated townhall to address any concerns from the SDSS community, and explicitly ask in the demographic surveys if members have read it.
  - External collaborators must explicitly agree to the Code of Conduct, alongside the Publication Policy.
  - Publicly published several guidelines and documents for SDSS and placed them in public repository and a subset on the SDSS public website, including inclusive chairing guidelines, how to run an inclusive telecon, accessible meetings guidelines, among others.[6]
- Efforts to increase SDSS membership from marginalized groups:
  - During the first few years of SDSS-IV, CPMS helped SDSS institutions in the United States run Research Experience for Undergraduates (REU) programs, where undergraduates used SDSS data and were part of the SDSS community for the summer (sadly this program ran out of funding).
  - CPMS and later COINS helped facilitate the Faculty And Student Teams (FAST) programs to fund and partner SDSS members with under-served communities. Professors from non-SDSS institutions could apply and be partnered with an SDSS mentor. Through FAST, students (typically from more marginalized groups) would become SDSS members and gain valuable research experience and connections. FAST has continued into SDSS-V.
- Encouragement of more inclusive behavior and attitudes within SDSS:
  - We have established and are reinforcing both formal and informal channels for addressing conflict, with an emphasis on contacting the SDSS Ombuds.
  - Normalization of some behaviors, such as displaying pronouns in an online display name, having discussions about inclusion in meetings/ communities, and giving space in agendas or meeting schedules.
  - Anecdotal evidence that our SDSS meetings are more welcoming and easy to participate in compared to other scientific meetings, and some quantitative data to this effect in our post-meeting surveys.
  - Normalizing the use of surveys to collect feedback from members rather than relying on impressions or anecdotes.

---

[4] https://www.sdss.org/collaboration/the-sloan-digital-sky-survey-code-of-conduct/
[5] For example, the Best Practices for an Inclusive Collaboration https://github.com/sdss/coins/blob/main/documents/md/best_practices.md
[6] The list of documents is maintained here: https://github.com/sdss/coins/tree/main/documents/md

    – Over the last years, general growth in sense of inclusion as measured in surveys (e.g., Figure 5.2).
    – Greater visibility of minorities, overall; anecdotally, members feel comfortable bringing up discussions of "social" issues that impact their communities.

Over the years, COINS has tried to overcome several re-occurring limitations, which we list below. The underlying theme is these shortcomings lead to a lack of resources and time needed to implement and enforce inclusive and accessible practices within SDSS and the community.

- COINS suffered from the perception of service work being inferior to scientific or technical work. The later tends to have more specific deliverables and deadlines. There is also the perception that service work is less helpful to progress one's career.
- While COINS' efforts became more recognized within SDSS, it is difficult to extend this to external institutions. This again ties into service work being perceived as less important. For SDSS service work, there is an additional step where work within SDSS may not be seen by the broader community. Lately we have been trying to raise the visibility of our work through publications and presentations to address this issue.
- COINS has top-level leadership support, but this can conflict with leading ground-up, grass roots initiatives. Being a committee that is mostly self-organized and responsive is a benefit for monitoring and helping change the SDSS climate. However it can lead to some complications, such as not having an independent budget for many of our activities (e.g., printing, prizes, tools, trainings). Since we did have leadership support, when we asked for funding for specific tasks it was warmly received, but it would be better if we did not have to ask so we could be more independent and responsive to the SDSS communities needs. Having the right balance between leadership support and being autonomous can be difficult to maintain.
- Despite being foundational to the existence of SDSS-IV and having broad support from top-level leadership, there were always small, but vocal, groups in opposition to COINS' efforts. This leads to more difficulty for those on the forefront of change, and often time is spent motivating why inclusive practices are important instead of implementing them.
- We need to curate and remind others of our presence in committees that are formed outside of the leadership, such as conference organizing committees. Because of this disconnect, COINS was often brought into conversations after critical decisions had been made (as discussed earlier). Thus, we had to adapt inclusive practices to a non-ideal environment, which more often than not, required intense, short bursts of attention.

Despite the shortfalls listed above, COINS has benefited SDSS greatly in its tenure in SDSS-IV. The challenge is maintaining this momentum within SDSS and beyond to ensure that inclusiveness is a central tenant in research efforts. Having support from SDSS leadership and the current cultural trend towards encouraging inclusive practices has helped COINS establish itself within SDSS and the astronomical community. We hope that others and other collaborations can build upon COINS' experience for a more diverse, equitable, and inclusive future.

References

Blanton, M., Anderson, S. F., Basu, S., et al. 2019, BAAS, 51, 196

Blanton, M. R., Bershady, M. A., Abolfathi, B., et al. 2017, AJ, 154, 28

Jones, A. M., Beaton, R. L., Cherinka, B. A., et al. 2023, PASP, 135, 124503

Kollmeier, J. A., Zasowski, G., Rix, H.-W., et al. 2017, arXiv:

Lucatello, S., & Diamond-Stanic, A. 2017, NatAs, 1, 0161

Lundgren, B., Kinemuchi, K., Zasowski, G., et al. 2015, PASP, 127, 776

York, D. G., Adelman, J., Anderson, John, E. J., et al. 2000, AJ, 120, 1579

AAS | IOP Astronomy

An Astronomical Inclusion Revolution
Advancing diversity, equity, and inclusion in professional astronomy and astrophysics
Dara Norman, Tim Sacco and Dorian Russell

Chapter 6

The Preparing for Astrophysics with LSST Program: Leveling the Playing Field

R A Street and F Bianco

6.1 Introduction

The NSF's Vera C. Rubin Observatory's Legacy Survey of Space and Time (LSST) will deliver a multi-band optical imaging survey of unprecedented depth, resolution, and temporal cadence across the whole of the Southern sky for 10 years, starting in 2024. As a result, LSST will be transformatory to most, if not all, fields in astrophysics, from exploring the origins of Dark Energy, to a census of Solar System small bodies, to mapping the Milky Way Galaxy and environs. It will reveal transient phenomena[1] of all kinds and on all timescales from hours to years. The full wealth of its science return is discussed in (Ivezic et al. 2019). The facility was accordingly recognized by the 2010 Decadal Survey[2] as the top ranked project, "The top rank accorded to LSST is a result of [...] its compelling science case and capacity to address so many of the science goals of this survey" (New Worlds, New Horizons in Astronomy & Astrophysics 2010).

Astronomers and data scientists from all backgrounds worldwide therefore have great interest in this flagship facility. Rubin data will be widely accessible, with no proprietary restrictions to any scientist in the USA and Chile as well as being open to many communities around the world who gained data access through in-kind contributions. The rest of the world will have access to the full data two years after its initial release. But not all researchers have equitable opportunities to explore these data. Long-standing imbalances in access to resources, knowledge, and opportunities, both in science and in society as a whole, combined with the pervasive

---

[1] Astronomical phenomena whose brightness or position is time dependent.
[2] A fuller discussion of the Decadal Survey is in Donahue's article in Chapter 15.

impacts of systemic biases, have resulted in a distribution of astronomers that is not representative of society more generally.[3]

Rubin (and other modern surveys) are at risk of compounding these issues through technological means. Rubin will produce 20 terabytes of imaging data *per night*, from which it is expected to generate up to millions of discoveries a day. The sheer volume and rate of delivery of these data products will preclude their exploration by the familiar tools of astronomy. Instead, a great deal of time, effort, and expense has been, and continues to be, invested in building "Big Data" software-based tools capable of handling these data. The ability to exploit the LSST data for science therefore depends on experience with and training in such advanced (and in some cases entirely new) tools, methods, and systems, and access to adequate computing resources.

Unlike many NASA missions and other large facilities, Rubin does not have a dedicated, funded science team. Rather, the NSF and DOE have committed to funding the construction of the observatory and its operation for the 10 yr duration of LSST. The astronomical community have self-organized into the Vera C. Rubin LSST Science Collaborations, of which eight are currently active, each dedicated to a different science topic. Only one of these Collaborations, the Dark Energy Science Collaboration (DESC; Bianco et al. 2019), receives direct federal support.[4]

This funding environment means that wealthy institutions have a systemic advantage, as they can make alternative funding sources available to support researchers, allowing them to begin preparations for science with LSST sooner. Frequently, the same set of institutions are able to provide advanced computing resources, software and IT training and support, as well as access to other observing facilities, which further advantages researchers at those institutions. This ability to lay vital groundwork for science in advance of the start of LSST creates a long-lasting productivity- and responsiveness-differential among institutions (which has a particularly severe impact on transient astronomy, as discussed in the next section). In the US, the institutions with the resources to do so disproportionately under-serve Black, Indigenous, and People of Color (BIPOC) communities.

Seeking the means to support the largely unfunded LSST Science Collaborations and help researchers to overcome these barriers led to the creation of the program, "Leveling the Playing Field: Preparing for Astrophysics with LSST", thanks to the generous support of the Heising-Simons Foundation.[5] This paper describes this program and its impacts on scientists preparing for LSST. Section 6.2 describes the Science Collaborations in more detail and their role in supporting researchers, while the different aspects of the program itself are presented in Section 6.3. We evaluate the experience gained in Section 6.4 and consider the outlook for the future.

---

[3] see for example https://www.aip.org/statistics/reports/demographics-and-intersectionality-astronomy
[4] DESC is partially supported by the Department of Energy DE-AC02-05CH11231.
[5] Grant 2021-2975.

6.2 Rubin Science Collaborations

The Rubin Science Collaborations are self-organized communities of researchers dedicated to different fields of astrophysics with LSST. Originally formed in 2006 with the goal of documenting the survey's scientific potential in the first LSST Science Book (LSST Science Collaboration 2009), there are eight currently active Science Collaborations (SCs) organized around science topics: Active Galactic Nuclei (AGN), Dark Energy (DESC), Informatics and Statistics, Galaxies, Solar System (SSSC), Stars, Milky Way and Local Volume (SMWLV), Strong Lensing (SLSC), and Transients and Variable Stars (TVS).

Membership of all of the SCs is international, including hundreds of scientists worldwide. The SCs provide scientific guidance to Rubin Observatory, for example, to develop scientific opportunities and on the impacts of survey design choices. The SCs act as a coordination hub for their respective communities, and a conduit for information both to and from researchers and Rubin Observatory. The size of the Rubin science community, and the common interest between SCs has resulted in many researchers contributing to more than one SC, bringing productive cross-fertilization of analysis and technical goals. The SCs are particularly valuable for junior career researchers to make contacts and gain exposure within their field, and can enable researchers new to Rubin to gain familiarity with the large scope of the project and access to opportunities. The eight SCs share an aspiration to be truly inclusive and equitable research environments. However, most of the SCs do not receive regular funding.

Repeated applications for funding for the SCs, both as an organization and through individual Principal Investigator (PI) programs, have encountered a marked reluctance of science panels to support preparatory science for a future survey in competition with projects producing results in the near-term. SC members are encouraged to seek funding independently. Some funding opportunities are provided by the Discovery Alliance[6]—a 503(c)(3) non-profit organization charged with obtaining philanthropic support for science (and formerly operations) leading up to and during the LSST, such as the Enabling Science grants[7] program and more recently the Catalyst Fellowships for post-doctoral scholars.[8] These opportunities continue to be an extremely valuable resource for the community, but the extent of the available programs pales in comparison with the scale of science that Rubin is capable of enabling. For example, the Catalyst program supports cohorts of ~5 Fellows at a time, and the typical scale of Enabling Science Grants is $20,000, which is small by federal standards.

Differential access to resources is particularly impactful in time-domain astronomy, which is at the heart of the TVS activities. The fleeting nature of transient phenomena, and the need to characterize them immediately after they are discovered, make it a technologically demanding science and has driven the development

---

[6] Formerly known as the LSST Corporation.
[7] https://www.lsstcorporation.org/enabling-science
[8] https://www.lsstcorporation.org/catalyst-fellowship/node/1

of alert brokers, Target and Observation Managers, and many other technologies for alert-based astrophysics. Only projects and investigators that are well prepared in advance of the start of the survey will be in a position to perform this follow-up for the largest possible sample of discoveries.

In the context of the Preparing for Astrophysics Program, the science areas of three of the SCs in particular overlapped those supported by the Heising-Simons Foundation:[9] TVS, SMWLV and SSSC. The co-chairs of all three participating SCs were invited to form the science advisory group for the program, which was administered through Las Cumbres Observatory.[10]

6.3 The Preparing for Astrophysics with LSST Program

The design philosophy of the Preparing for Astrophysics Program[11] was to identify, with the mission of overcoming, the barriers that prevent researchers from participating in Rubin research, and to lay the groundwork of partnerships that would enable the involvement of communities that have previously been excluded. The program was designed explicitly to mitigate imbalances observed within the TVS, SSSC, and SMWLV collaborations. The original proposal was limited to a request for support for publications, but working in close connection with Heising-Simons Foundation representatives, it was expanded to a more ambitious and comprehensive program consisting of four elements. These focused on the primary avenues by which the SCs could make an immediate impact on overcoming barriers to participation: Financial support for publications, financial support for workshops, computing and software infrastructure and training, and kickstarter grants to foster new collaborations. Each of these elements is described in the following sections, and more information can be found on the program's website.[12] One defining factor influencing the design of this program was its short 1-yr duration, which was a condition of the funding.

6.3.1 Kickstarter Grants Program

Prior to this program, discussions with many SC members highlighted a common problem: they would like to dedicate more time to LSST preparations, but their existing funding demanded they focus on other commitments. Making grants available for LSST research was therefore the highest priority for this program.

The funds available for this program element were limited to about $720,000. The combined membership of the participating SC's membership is about 700 people. Thus, the question became how to enable the maximum scientific return, ensure equitable access to the opportunity, and support a very diverse range of science in a fair manner. While there is a need for an organization, like the SCs, to obtain grants that can be used to hire postdocs and development of large scale Rubin projects such

---

[9] https://www.hsfoundation.org/

[10] https://lco.global/

[11] https://lsst-sci-prep.github.io/

[12] https://lsst-sci-prep.github.io

as graduate theses, considering the size of this grant and its short timescale, we decided that the most impact could be achieved by supporting many small grants enabling the involvement of new and recent LSST SC members and recruiting into ever broader communities.

We opted to offer members the opportunity to apply for small grants through a unique call for proposals that incentivized "equitable" research partnerships. Targeting BIPOC communities as populations traditionally marginalized in STEM in the USA has been a long standing goal of the TVS SC. The international membership of the SCs benefitted from a broad definition of what programs aimed at "inclusion" may look like and who it may involve and support. The call for proposals explicitly highlighted both scientific excellence and equitable opportunity in science as the twin goals of the program. Two options were offered: team/individual grant proposals could request up to (US)$20,000 but partnership proposals could request up to $30,000. This category of proposals had to be designed to overcome barriers to entry to Rubin research and/or form meaningful partnerships between institutions that are new to Rubin and those with established Rubin programs.

A template proposal form was provided which included optional sections where teams could highlight any personal circumstances that particularly inhibited their ability to participate in Rubin research. All proposals were peer reviewed by an eight-person panel of science specialists, Rubin Observatory representatives, and advocates for diversity and equity. Proposals that sought to lay the groundwork for research collaborations that could be sustained long-term were particularly encouraged, especially involving communities that have historically been excluded from astronomy.

A total of 35 proposals were received. Over 40% of proposals involved international teams, and 50% of proposals in the Partnership category were led by female Principal Investigators, 52% in the Individual/Team category. A total of 126 researchers directly collaborated on the proposals, but many more were actively involved in the projects, especially since several proposals recruited students and hosted workshops. Among these proposals we identified projects that fostered our DEI goals of supporting the development of Rubin-related research activities at Minority Serving Institutions (MSI) including Hispanic Serving Institutions and Historically Black Colleges and Universities (HBCUs); supporting researchers from South-American countries that had recently entered the Rubin ecosystem (Brazil, Argentina); supporting PIs and researchers in vulnerable stages of their career or with care-giving responsibilities that competed with their research time (see Figure 6.1).

At least two programs were explicitly aimed at inclusion of differently-abled scientists (3 and 4 below). One program (3) developed an interdisciplinary partnership between an R1 and a Historically Black College and University, which involved astronomers and musicians and one program supported mentorship relationships between established SC members and incoming faculty-student pairs from Minority Serving Institutions (1).

Many of the projects funded Rubin research by students and post-doctoral researchers, as well as enabling staff to dedicate time to Rubin research. This is particularly important for those in non-permanent positions, and those who needed

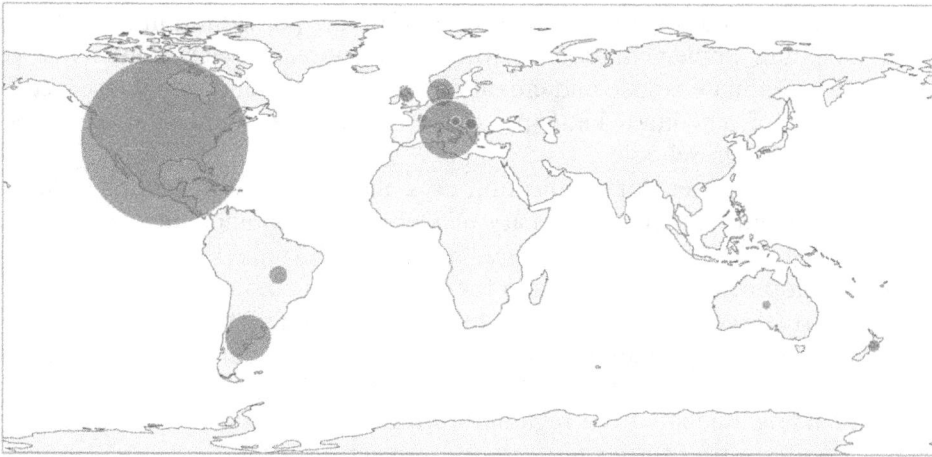

Figure 6.1. Map of the locations of participants in the Kickstarter Grants. The diameter of each circle is proportional to the number of participants in each country.

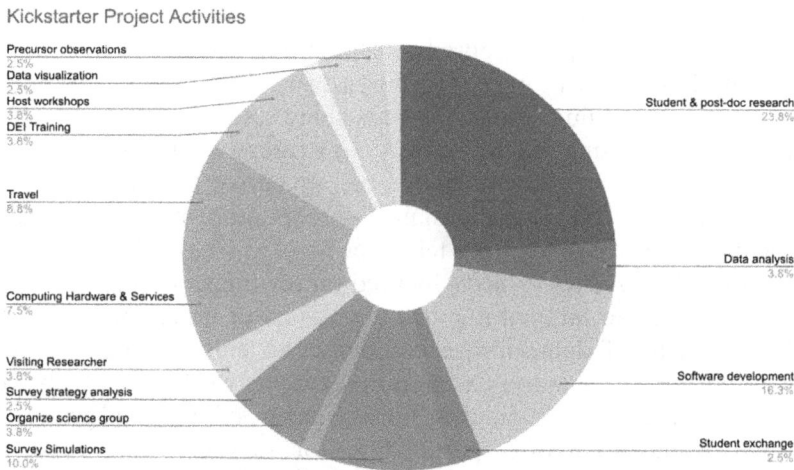

Figure 6.2. Summary of activities undertaken by Kickstarter Projects.

to buy-out of teaching requirements to make time for research. This program explicitly included support for child/dependent care costs to help researchers with family commitments return to research, a feature that was especially valued as the world recovered from the impacts of pandemic-related school closures.

Figure 6.2 illustrates the range of activities undertaken, while the lists below provide greater context for those activities.

Kickstarter Programs focused on inclusion and equity in STEM:

1. A program supported mentorship relationships between four established SC members and four incoming faculty-student pairs from Minority Serving Institutions

2. A collaboration with representatives of native population in New Zealand including implementation of Maori astronomical science in a first-year undergraduate course and creation of astronomy educational resources for schools from kura kaupapa (Maori language immersion preschools) through high schools
3. The development of sonification as a means of accessing LSST data—a partnership between University of Delaware (R1) and Lincoln University (HBCU)[13]
4. Development of 3D rendering for Rubin stellar models and data
5. A virtual workshop to provide bystander training[14] to SC members
6. Participation by Rubin researchers in a workshop during the International Day of Women and Girls in Science (Feb. 11, 2022 in Palermo, Italy) including students from regional institutions.
   Kickstarter Projects focused on outreach
7. Development of the Trailblazer database to track astronomical data affected by commercial satellite trails.
8. Outreach to high school students in Italy to raise awareness of Rubin as well as equity in science.
   Kickstarter Projects focused on students' activities
9. International student exchange program between universities in the USA and France enabling students to contribute software and statistical analysis towards optimizing Rubin's observing strategy for transient detection
10. Telescope time for students to conduct precursor observations of Globular Clusters at the Anglo-Australian Telescope and of planned Rubin Deep Fields at the Mt. John 1.8m telescope in New Zealand.
    Other Kickstarter Projects focused on revitalizing several science groups where work had faltered due to the pressure of other commitments.
11. Procurement of data storage equipment to enhance regional computing facilities for Serbian and Croatian researchers working on Rubin research.
12. Supported active participation in a number of working groups from all three participating SCs, particularly for survey strategy and the Data Preview 0 program (designed to enable the community to build familiarity with the Rubin data products and the Rubin Science Platform interface)
13. Stimulated the rejuvenation of some scientific working groups, in particular the TVS group for supernovae and pulsating stars, by providing stipends to those in leadership positions. This allowed them to dedicate time to the organization of the groups, as well as contribute to specific activities, including contributions to the TVS Science Roadmap, liaising with other

[13] This program and additional funds from Rubin and LSSTC supported the participation of the Lincoln University faculty-student pair into the annual Rubin all-hands meeting https://project.lsst.org/meetings/rubin2022/%3Cfront%3E
[14] Bystander Training is intended to give those who witness harassment or bullying happening the real-time tools to intervene in order to stop the activity.

Rubin Science Collaborations, and participating in Rubin's annual Project and Community Workshop.

14. Development of several open-source software and documentation for key investigations into Rubin data on Solar System objects, that will enable broader participation.

One of the successes of this program element was to stimulate participation in the SCs by many researchers from across Brazil and Argentina, nations which have not previously been represented. For example, one project team reported that a nationwide selection of students garnered 97 applicants, from whom six were selected to work on stellar systems in the Milky Way. The diverse communities and range of research that these projects were able to foster is a gratifying outcome to this program in the short term.

6.3.2 Workshops

Conferences, meetings and workshops are essential to cross-fertilize and inspire innovative research, to share results and for junior career researchers to network. The recent and ongoing global travel restrictions due to the SARS-CoV-19 pandemic severely hindered these interactions but also accelerated the acceptance of online meeting platforms as an alternative. The benefit of platforms such as Zoom, Microsoft Teams, Gather and others is that they enable researchers to participate in meetings from around the world, without the financial burden, or environmental impact, of travel. They also enable greater participation by those with dependant-care responsibilities. That said, they can require substantial internet bandwidth, which limits their accessibility in some regions of the US, and in many other countries. We found that if meetings are designed with remote participation in mind from the outset, they can be rewarding for all concerned. As pandemic restrictions on travel were still in force when this program was conceived, we planned all of our workshops to be fully virtual, but these logistical considerations hold true for hybrid (in person+virtual) meetings. None of the program workshops had a registration fee, and we made use of Zoom for plenary sessions and Gather for interactive breakout groups. Channels were set up for each workshop on the LSSTC Slack team to enable real-time discussions.

Table 6.1 summarizes the topics of the workshops organized as part of this program. Efforts were made to advertise the first workshop as widely as possible, in order to attract researchers who may not have felt able to participate in Rubin research previously. Adverts were distributed through large organizations in astronomy, both within the US (e.g., the American Astronomical Society, the US National Society of Black Physicists) and internationally (including the International Astronomical Union (IAU), the European Astronomical Society, the African Astronomical Society and the Royal Astronomical Society), as well as through mailing list to all members of the participating Rubin SCs.

Lessons learned and best practices: Given the wide geographic distribution of the SC members, we assumed at the outset that many people would participate in

Table 6.1. Summary of the Workshops Organized by this Program

Workshop	Dates	Purpose
Kickstarter Workshop	September 8 and 9, 2021	Introduce the Kickstarter research program to the community, foster collaborations, help new members to get involved, and answer questions about the program.
TVS Software Workshop II	May 4 and 5, 2022	Coordinate the development of software for TVS research, and gather community input.
Software Development Skills Workshops	May 23–27, 2022 July 11–15, 2022	This series of workshops was designed to provide training in software development skills going beyond the level typically taught in astronomy.

workshops asynchronously. A participant's timezone can be established when they register for the workshop, enabling us to time sessions during working hours for the majority of attendees. Sessions lasted no more than 3–4 h with breaks of at least 30 min to minimize "Zoom fatigue". Later sessions reviewed the content presented earlier. Working groups were structured regionally when appropriate (e.g., for group training exercises) to enable real-time interactions, even if they took place offset in time from other groups at the meeting. At training workshops, we found it beneficial to recruit instructors or assistants on different continents, so that someone was always available to answer questions. All sessions nominated distinct people in the roles of session chair, scribe for recording the notes, and moderator to monitor the Zoom chat and Slack channel for questions.

Shared cloud-based documents were used to record notes of all sessions, and for all meeting content, training materials, and products from the meeting. Participants in timezones substantially offset from the meeting sessions were invited to contribute to the shared documents between sessions, and this was found to be beneficial for those in regions of low- or intermittent-bandwidth. All meeting content and training materials were posted online, ideally in advance, to allow for downloading when bandwidth permitted. Recordings of each day's sessions were posted to YouTube as soon as possible after each session, to enable those in different timezones to keep pace with proceedings. We note that large international scientific collaborations represent a social network which can be a powerful way to make connections. For example, during this program, the PI was invited to speak at a virtual workshop organized by members of TVS in Serbia (A. Kovacevic and D. Ilic) in partnership with their long-standing collaborators in Ethiopia. Although the Serbian program was over halfway through by the time of the workshop, the talk provided an opportunity to advertise the program's remaining events more widely to researchers in Ethiopia and elsewhere in Africa. This led to several of the African researchers participating in our Software Development Skills Workshops.

6.3.3 Computing and Software

The volume and rate of delivery of LSST data make access to substantial computing resources indispensable if research goals are to be achieved. It also mandates familiarity with the cutting-edge software tools necessary to make use of those computing resources. These include Cloud-based computing and storage, remote data-access portals such as the Rubin Science Platform (Dubois-Felsmann et al. 2019), and industry standard protocols like Apache Kafka, which is used to disseminate the Rubin alert stream.

However, the training available to astronomers in these topics is extremely patchy. Most receive some formal instruction in the basics of a programming language (usually Python, C/C++, Fortran), but little advanced training is available. In particular, few are trained in software engineering skills, such as program design patterns, or unit testing, that produce robust, verified code that can be maintained over the long term. Although these skills would lead to greater confidence and reliability of scientific results, astronomers are often disincentivized to seek training in these areas as this would take time away from their topic of research, and can be expensive. Once again, larger institutions that can employ software engineers and IT specialists to work with their researchers have an advantage over places where such resources are not easily available.

Happily, the Rubin science community has a number of options to overcome the lack of computing resources. Rubin Observatory will dedicate 10% of its computing to community science, which will be met by providing access to the survey data products through the Rubin Science Platform.

Access to this platform will be available to all Rubin Data Rights holders, and is currently ramping up. In addition, the NSF's Optical and Infrared Research Laboratory, NOIRLab, has launched the Astro DataLab platform. Offering similar browser-based tools, but access to the large data holdings of the DataLab archive, this facility is open to all astronomers. Similar portals are planned for a number of International Data Access Centers, which are currently in development.

Shortly after this program began, we learned of the new LINCC Frameworks Program,[15] which includes providing access to a large computing cluster at the DiRAC Center at the University of Washington. We were pleased to collaborate with this program, and made sure all participants in our program were made aware of these resources.

These resources liberated some program funds that were originally earmarked to secure computational resources to provide access to training in advanced software design and engineering skills. Firstly, we offered members the opportunity to complete the industry-recognized IEEE course in Recommended Practice for Software Design Descriptions. Ten members took this online course. Secondly, we organized access to two virtual workshops for the "Intermediate Re search Software Development in Python" course developed by Software Carpentry, and presented by the Software Sustainability Institute. Forty-six researchers attended from Argentina,

---

[15] https://www.lsstcorporation.org/lincc/frameworks

Croatia, Ethiopia, Germany, Hungary, India, Israel, Italy, New Zealand, Serbia, South Africa, UK, USA.

6.3.4 Publications

The cost of publishing a single paper in a refereed journal can run to several thousand US dollars, and researchers face strong career pressures to publish in high-impact journals. While some institutions provide funding to their staff for this purpose, many do not, particularly outside the US. This disparity was the original motivation for our seeking funding to support the publication of research into LSST's survey strategy by TVS members.

Furthermore, at this stage the LSST-related publications focus on science-reparation and science driven advocacy, rather than discovery, and publications of this nature typically encounter more obstacles in the refereeing process than publications that can claim a science discovery. Junior scientists and scientists from less known institutions are at an inherently higher career risk when they engage in work that does not directly lead to publications. Yet the work done by the SC members at this stage, including participating in the refinement of the LSST survey strategy and roadmapping the science path for their respective SCs, is critical and time consuming.

Prior to the start of our Program we had worked with the Institute of Physics (IOP) editors to secure a friendly venue sympathetic to science preparation related work. At time of writing, papers on LSST survey strategy research are being collected into a dedicated focus issue of the Astrophysical Journal's Supplement Series.[16] We then sought financial support for these publications. Our Program has funded 12 papers for this focus issue so far (Andreoni et al. 2022; Bellm et al. 2022; Bonito et al. 2022; Dal Tio et al. 2022; Di Criscienzo et al. 2022; Hernitschek & Stassun 2022; Kovačević et al. 2022; Li et al. 2022; Prisinzano et al. 2022; Raiteri et al. 2022; Schwamb et al. 2022; Street et al. 2022). Since survey strategy is clearly critical to all science with LSST, members of all eight SCs were invited to apply for support through this program, and the papers submitted reflect this diversity of science.

We also earmarked funds to enable the publication of "science Roadmaps" by the TVS (Hambleton et al. 2023) and SMWLV SCs (SSSC having published theirs previously (Schwamb et al. 2018)). Roadmaps are very substantial papers, critical to set a path for a SC, jointly written by many members, which describe the scientific goals of a collaboration and the path to achieve those goals. Roadmaps are often highly cited documents, but can extend to over 100 pages and are therefore costly to publish.

Funds were also allocated to the publication of any other LSST-related research by any member of the three participating SCs.

Lessons learned: It became evident throughout the Program that one year is too short of a timeline to effectively support publications given the lengthy process of

---

[16] https://iopscience.iop.org/journal/0067-0049/page/rubin_cadence

writing, submitting, and revising a paper all the way to publication (even for work that was typically already underway at the time when the award was received). Particularly, scientists with heavy teaching or service duties, or because their salaries are closely tied to specific deliverables, could not shift the scope of their work to focus on rapid turn-around publications. With this in mind, and considering that the Rubin Observatory's process for determining the survey strategy is ongoing (Bianco et al. 2022), as is the SCs work on evaluating proposed surveys, we extended the timeline for requesting and receiving financial support for publication through 2022–2023.

6.4 Experience Gained and Lessons Learned

While this program met its immediate goals of supporting a diverse research community, and enabled a wide variety of LSST-preparatory activities, its long-term impact will depend upon the continued availability of funding. The hope is that, with the start of Rubin Observatory operations in the near future, government funding sources will become available. For example: a concrete risk is that the new collaborations enabled by these Kickstarter projects, including the involvement of scientists from minority serving institutions, will not be sustainable in the absence of funds for teaching buy-outs and to support student internships.

Objective metrics for evaluating the impact of this program are somewhat difficult to design. For example, we can measure the number of researchers who have become active contributors to the SCs, or the number of students and post-doctoral researchers funded. More difficult to measure is the impact of this program on the involvement of Black, Indigenous, and People of Color in Rubin. While some Kickstarter projects clearly identified their support for BIPOC communities, and some highlighted support for under-resourced communities, not all did so. To improve our insight into the impact of our activities in the future, we are in the process of holding a census of the Rubin science community, with the goal of repeating such surveys at regular intervals.

As the world returns to travel post-pandemic lockdown, the value of in-person meetings is still without parallel for the formation of new and serendipitous collaborations and networking. Nevertheless, well thought-out provision for virtual attendance in all meetings has been clearly demonstrated to overcome a number of barriers to participation, and we hope that future meetings will continue this practice. The availability and rapid turnaround of the Kickstarter Grants was advantageous during the pandemic, particularly for those in soft money positions, and to provide summer student internships when other options for employment were suddenly curtailed. The long tail of ongoing COVID infections has been one of the leading causes of delays in project completion, an indication that our recovery from the disruption is ongoing.

Feedback from attendees of the software training courses was uniformly positive, but offered constructive suggestions on the format of the presentations. More real-time code demonstrations and direct interactions were felt to be beneficial. Informal polls suggested that the community was most interested in intermediate-advanced level training, and the view was expressed that universities should be held responsible for providing at least a basic level of tuition. Feedback was limited, and perspectives may vary worldwide. We did note that it can be difficult for astronomers to justify attending courses on "abstract" software engineering techniques to their institutions, unless their institution already has an interest in software development. This can lead to low take-up. A more impactful approach might be to provide access to training in topics or services with clear direct application to research.

References

Ivezic, Z., Kahn, S., Tyson, J. A., Abel, B., Acosta, E., et al. 2019, ApJ, 873, 111

National Academies of Sciences, Engineering, and Medicine 2010, New Worlds, New Horizons in Astronomy and Astrophysics. (Washington, DC: The National Academies Press) https://doi.org/10.17226/12951.

Bianco, F., Banerji, M., Blum, R., Bochanski, J., Brandt, W. N., et al. 2019, BAAS, 51, 185

LSST Science Collaboration Abell, P. A., Allison, J., Anderson, S. F., Andrew, J. R., et al. 2009, LSST Science Book, Version 2.0, arXiv e-prints, arXiv:0912.0201

Dubois-Felsmann, G., Economou, F., Lim, K., Mueller, F., Pietrowicz, S. R., & Wu, X. 2019, LDM-542 Science Platform Design https://ldm-542.lsst.io

Hambleton, K. M., Bianco, F. B., Street, R., Bell, K., Buckley, D., et al. 2023, PASP, 135, 1052

Schwamb, M. E., Jones, R. L., Chesley, S. R., Fitzsimmons, A., Fraser, W. C., et al. 2018, Large Synoptic Survey Telescope Solar System Science Roadmap, arXiv e-prints, arXiv:1802.01783

Bianco, F. B., Ivezić, Z., Jones, R. L., Graham, M. L., Marshall, P., et al. 2022, ApJS, 258, 1

Li, X., Ragosta, F., Clarkson, W. I., & Bianco, F. 2022, ApJS, 258, 2

Raiteri, C. M., Carnerero, M. I., Balmaverde, B., Bellm, E. C., Carkson, W., et al. 2022, ApJS, 258, 3

Andreoni, I., Coughlin, M. W., Almualla, M., Bellm, E. C., Bianco, F. B., et al. 2022, ApJS, 258, 5

Hernitschek, N., & Stassun, K. G. 2022, ApJS, 258, 4

Bellm, E. C., Burke, C. J., Coughlin, M. W., Andreoni, I., Raiteri, C. M., & Bonito, R. 2022, ApJS, 258, 13

Kovačević, A. B., Radović, V., Ilić, D., Popović, L. C., Assef, R. J., et al. 2022, ApJS, 262, 49

Dal Tio, P., Pastorelli, G., Mazzi, A., Trabucchi, M., Costa, G., et al. 2022, ApJS, 262, 22

Bonito, R., Venuti, L., Ustamujic, S., Yoachim, P., Street, R. A., et al. 2022, ApJS, 265, 27

Prisinzano, L., Bonito, R., Mazzi, A., Damiani, F., Ustamujic, S., et al. 2022, ApJS, 265, 39

Di Criscienzo, M., Leccia, S., Braga, V., Musella, I., Bono, G., et al. 2022, ApJS, 265, 41

Schwamb, M. E., Jones, R. L., Yoachim, P., Volk, K., Dorsey, R. C., et al. 2022, ApJS, 266, 22

Street, R. A., Li, X., Khakpash, S., Bellm, E., Girardi, L., et al. 2022, ApJS, 267, 22

AAS | IOP Astronomy

An Astronomical Inclusion Revolution
Advancing Diversity, Equity, and Inclusion in Professional Astronomy and Astrophysics
Dara Norman, Tim Sacco and Dorian Russell

Chapter 7

Anti-Black Racism Workshop during the Vera C. Rubin Observatory Virtual 2021 Project and Community Workshop

Andrés A Plazas Malagón, Federica Bianco, Ranpal Gill, Robert D Blum, Rosaria (Sara) Bonito, Wil O'Mullane, Alsyha Shugart, Rachel Street and Aprajita Verma

7.1 Introduction

The Vera C. Rubin Observatory (Rubin) was ranked as the highest priority ground-based astronomical facility in the 2010 National Academies of Sciences, Engineering, and Medicine Decadal Survey of Astronomy and Astrophysics (Astro2010).[1] The same report recognized the lack of progress in increasing the number of minorities in astronomy and the work that is still needed to eliminate gender gaps in our field. A decade later, the next iteration of this decadal report[2] placed a more explicit emphasis in how Equity, Diversity, and Inclusion (EDI) are fundamental for the foundations of the profession. Understanding racism as structural and systemic, rather than being an individual and intentional occurrence, can prompt organizations to confront the root causes of racial disparities and lack of progress with EDI efforts. EDI is important for the Rubin Observatory not just on an ethical level, but also for the potential to maximize the science return of the project by (among other things) broadening the talent pool.[3] During Rubin's 2021 virtual Project and Community Workshop (PCW), the annual Rubin community based meeting, we proposed the realization of an anti-Black racism workshop

---

[1] *New Worlds, New Horizons in Astronomy and Astrophysics*, see Chapter 6 for more details about the Astronomy and Astrophysics Decadal Surveys.
[2] *Pathways to Discovery in Astronomy and Astrophysics for the 2020s.*
[3] See, for example, Page, S. E. "The Difference: How the Power of Diversity Creates Better Groups, Firms, Schools, and Societies." 2008, and Nature editorial "Science benefits from diversity", 2018.

facilitated by an external organization (*"The BIPOC Project"*).[4] The initial plan proposed having a workshop on systemic racism and its relation to EDI, however, triggered by the events that took place in 2020 in the USA and at an international level in the aftermath of the murder of George Floyd, the Black Lives Matter movement and the Strike for Black Lives on 19 June 2020,[5] the final workshop was more specifically focused to address anti-Black racism. Within what we refer to as the "Rubin Ecosystem",[6] it is clear that there is work to do to improve EDI and to address and dismantle systemic racism, in particular, anti-Black racism. The number of Black and African American people in the scientific and technical Rubin community is unlikely to surpass 1% if attention is not paid to this concern. Currently, there are no official metrics or statistics to track such Rubin workforce demographics with any accuracy; this is itself a problem. To create a community that is naturally inclusive and equitable, sustainable systems to achieve that goal need to be implemented to ensure that members of groups traditionally marginalized from STEM (Science, Technology, Engineering, and Math) and astronomy feel welcome and supported, thus increasing their retention in astronomy. The goal of the workshop was to educate a representative group of Rubin members, including senior leadership and people in decision-making positions, across the Rubin organization so that inroads could be made to put in place systems and processes that lead to positive change. After the virtual workshop participants were expected to "return" to their institution/collaboration and find ways to make actionable change for incremental progress that would eventually address the underrepresentation of Black people in organizations associated with Rubin Observatory. Given the predominant role of Rubin as one of the most important and visible astronomical projects of the decade, we hope that this also sets an example in the astronomical community in general.

7.2 The Rubin Ecosystem and Structure of the Workshop

The Rubin Observatory is a large organization. The Rubin Observatory includes several hundred employees that work on various aspects of Rubin Construction Project and Rubin Operations. These are separate, although overlapping, organizations with teams that focus on different areas of the project such as Data Management, Commissioning, etc. In total we identified eight subsystems within the Rubin Ecosystem. This staff includes Rubin's Chilean employees working in both the US and Chile.

There is also the community that is interested in generating science from the upcoming data from Rubin's Legacy Survey of Space and Time (LSST).

---

[4] https://www.thebipocproject.org/

[5] https://m4bl.org/events/strike-for-black-lives/, https://www.particlesforjustice.org/, https://www.shutdown-stem.com/

[6] The Rubin ecosystem is formed by funding agencies and institutions, the Rubin Construction Project (formerly known as LSST Construction project) which includes engineers, scientists who will operate the telescopes, international collaborations of scientists, etc A detailed structural map can be found here. https://noirlab.edu/public/images/rubin-rubin-ecosystem/

This includes thousands of people and many of these researchers are organized into the eight LSST Science Collaborations, which are supported by the LSST Corporation, a private non-profit engaging in fundraising. Both these organizations are external to Rubin, but they are key elements of the Rubin community and ecosystem.

A typical annual meeting, called the Project Community workshop (PCW), includes more than 500 participants from all parts of the Rubin EcoSystem.

7.2.1 Workshop Planning

Initiatives aimed at improving climate and equity from the inside, such as committees and discussion groups, are common both in academic STEM fields and in Rubin.[7] However, these initiatives, as well-intentioned as they are, often lack strategic goals and risk being slow or even completely ineffective at enacting change. With many equity issues to address in academic STEM, in most cases, these groups have little time and, for the most part, no resources for (self)-education and training. We decided that education was key to enact change and planned to recruit a professional team to facilitate training. The implementation of this required securing funds: the cost of such training is not trivial, typically running in the several thousands of dollars. We took advantage of a regular funding call for proposals issued by the LSST Corporation. We submitted a proposal to the LSSTC 2020 Enabling Science call, expecting to run in-person anti-racism training sessions at the 2021 Project Community Workshop, but when the PCW 2020 was moved to virtual format, the final decision was made to conduct the workshop remotely.

To plan the proposal we had contacted an organization ("*Crossroads*"[8]) that at the time organized anti-racism training for academic communities and based the budget on their quote. The timing was peculiar: our proposal was submitted before the start of events that catalyzed anti-racism and anti-Black racism, especially in the US. During the US "racial reckoning" triggered by the COVID-19 pandemic and the murder of George Floyd, organizations such as *Crossroads*—the organization we had initially contacted—and *The BIPOC Project*[9]—the organization we finally selected—saw a spike in requests. This led to an increase in the cost of this facilitated workshop, and more competition to recruit a facilitators team for the dates of the PCW, which were fixed. Several organizations, including *Crossroads*, had by 2021 reorganized their training offerings and were no longer providing small workshops such as the one we had planned. The cost also increased but the Rubin Observatory Leadership agreed to contribute to the expenses. However, the changing political

---

[7] Thomas et al. 2022: *Creating an inclusive and diverse environment at Vera C. Rubin Observatory* https://doi.org/10.1117/12.2630499

[8] https://crossroadsantiracism.org/

[9] https://www.thebipocproject.org/ Other organizations contacted included "Overcoming Racism" (https://www.overcomeracism.com/) and "Diversity Talks" (https://www.diversitytalkspd.com/)

landscape of 2020 meant that we had to be aware of the ability to use federal funding for the activity.[10] Ultimately, with a change of federal administration, we were able to use some federal funding.[11]

7.2.2 Selection of Facilitating Organization

In order to identify an organization to facilitate this workshop, we initially consulted with colleagues in different institutions and performed our own research in the literature and the internet in general. In the end, when selecting the facilitating organization we considered the following:

- o **Cost**: we had a hard limit to the budget. We made that clear to the organizations we contacted from the start.
- o **Timing**: our sessions were tied to the dates of the Rubin PCW with no flexibility beyond the PCW week. Furthermore, the meeting was offered online to an international community with attendees from as far west as Hawai'i and as far east as East Asia and Oceania. The live time was restricted to 10AM-4PM Eastern (USA) to limit the strain on participants from non-US time zones.
- o **Focus on anti-racism over broader equity and EDI topics**—to ensure effective-ness we wanted to ensure that this workshop was focused and did not generalize EDI issues. We had already chosen the focus to be anti-racism and further narrowed the scope to be anti-Black racism.
- o Given the scope, we decided it was critical that the **leadership of the facilitating organization included Black, Indigenous, People of Color (BIPOC) and that BIPOC facilitators would be visible to the participants**. This helped us to ensure that the recommendations we would be receiving would come from personal experience and knowledge, not just from academic research. The choice also allowed us to support a BIPOC owned and run business.
- o We ensured that the organizations had **experience with academic environ-ments, ideally STEM**. Given the cost, these facilitated workshops are more commonly held in the corporate world than in academia. We believe that those in academia are, in general, often exposed to equity concerns and issues, yet inequities, biases, and discrimination are still pervasive, and even more so in STEM. In addition, we acknowledge that STEM training and the tradi-tional STEM environments tend to be competitive and even aggressive; while EDI issues are more significant in STEM than in other Academic environ-ments, the personalities in the STEM networks tend to be analytical and skeptical of outsiders' approaches.

---

[10] https://www.npr.org/2020/09/05/910053496/trump-tells-agencies-to-end-trainings-on-white-privilege-and-critical-race-theor

[11] https://law.ucla.edu/news/biden-reverses-trump-executive-order-banning-diversity-training

○ Rubin is an international community and this exposes a significant issue in EDI work: equity, inclusivity, and racism issues may look different in different cultural contexts.[12] **We asked the organizations to describe how they would approach training with an international audience**. Ideally we wanted to ensure that the organization had prior international experience, but we found that very few such organizations work across international boundaries. We relied on a commitment of the organizers to prepare material that explicitly supported a multi-cultural, international audience.

During the selection process, we had several phone and video calls with representatives of the following candidate organizations: *Crossroads, Overcoming Racism, Diversity Talks, and The BIPOC Project*. Based on the criteria and constraints described above, in the end we selected *"The BIPOC Project"* to facilitate the following workshop at Rubin's 2021 virtual PCW:[13]

Building Black Power: Dismantling Anti-Blackness in Our Institutions and Movements—This experiential gathering tackles the root causes of anti-Black racism and its cultural influences within our institutions, movements and communities, and offers a framework for unlearning anti-Blackness and developing a pro-Black stance within anti-racist practice among BIPOC folks.

7.2.3 Selection of Participants and Leadership Involvement

The original proposal had envisioned an in-person training opportunity with the expectation to include between 100–150 participants. This number was based on the participation in structured, but not facilitated, EDI sessions held in previous years and on the expected number of participants. We did not initially anticipate having to select participants. However, the modification of the original plan described above and the maximum number of participants set by *The BIPOC Project* to maximize the effectiveness of training (based on their experience with prior events[14]), we were faced with the possibility of having to select participants among applicants. We created a sign-up sheet for the workshop within the PCW registration to assess potential interest. We indicated in the initial sign up and iterated for clarity by email that we expected full participation to the entire workshop, which was held over two days in parallel with scientific sessions at the PCW. Further, we explicitly recruited representatives of the Rubin communities to ensure the entire Rubin ecosystem would be exposed to this training. In addition, we recruited members of the Rubin Science Advisory Committee, the official body advising Rubin Observatory on scientific decisions; while the scope of their recommendations are primarily scientific, or science-motivated, we recognize that bias and racism are inextricably linked to

---

[12] In many cultures and countries, darker skin is also a basis for discrimination.

[13] https://www.thebipocproject.org/what-we-do

[14] *The BIPOC Project* organizes workshops for up to 60 Participants with two two-hour sessions over consecutive days.

our way of thinking and reflecting on the world, including our scientific thinking and science-driven prioritization.

Finally, we wanted to ensure participation by the organizations that manage and govern Rubin. Change can come from a grassroots level as has mostly been the case with the Rubin Observatory EDI initiatives. It is recognized that commitment from leadership sends a strong message to the rest of the organization. While the workshop was oversubscribed according to the interest expressed during registration to the PCW, a number of Rubin Leaders, both current and future, were explicitly invited to participate and approximately 80% attended. It was invaluable to have leadership involved because it allowed subsequent conversations to take place with those individuals using language, terms, and ideas that were learned in the workshop. We invited representatives from Rubin's main funding agencies in the USA, the National Science Foundation (NSF), and the US Department of Energy (DOE), as well as the leadership of the organizations that manage Rubin: the Association of Universities for Research in Astronomy (AURA), the NSF's National Optical-Infrared Astronomy Research Laboratory (NOIRLab) and DOE's SLAC National Accelerator Laboratory (SLAC).

In total we received 74 applications and obtained permission from *The BIPOC Project* to extend the participants' limit to this number so that we did not have to reject any applicants.

We monitored the demographic profile of our participants by asking questions in the application form including nationality and their seniority level, defined in three tiers: junior (students and postdoctoral scholars), mid-career (senior postdocs, scientists, junior faculty), and senior (faculty starting at associate level, senior scientists). We did not ask participants to identify their gender or ethnicity/race. In retrospect we realized monitoring the fraction of People of Color (POC) signed up for the workshop was particularly important. The organizers browsed the list of participants to provide additional race or ethnicity information for those whom they had prior knowledge and assigned them to two groups: POC[15] and non-POC in order to facilitate broad discussion in small groups (Figure 7.1).

7.3 Workshop Sessions

Both sessions of the two day workshop were facilitated by *The BIPOC Project*. The first session consisted of an introduction to the workshop and to fundamental concepts, definitions, and shared vocabulary. The desired outcomes from the workshop were as follows:

- ○ *"A shared language and analysis for understanding and addressing white supremacy and anti-Black racism as foundational to advancing diversity, equity, and inclusion within your community"*
- ○ *"Deeper awareness of the ways in which anti-Black racism shows up in ourselves, in society, and in our organizations"*
- ○ *"Initial steps to undo anti-Black racism, advance racial equity, and build a pro-Black scientific future"*

---

[15] We defined Person of Color (POC) as anyone who is not white.

Participants' Seniority

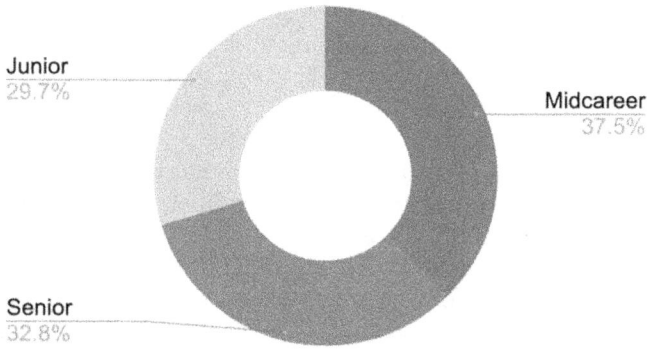

Junior
29.7%

Midcareer
37.5%

Senior
32.8%

Participants' Nationalities

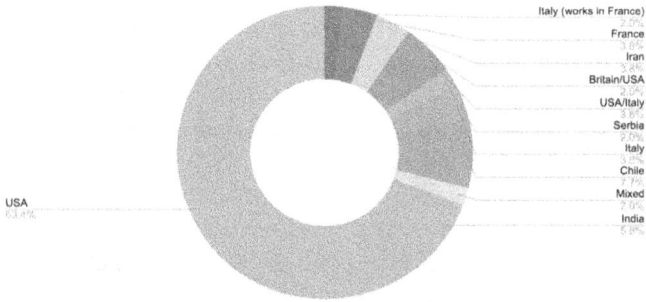

Italy (works in France)
2.0%
France
3.0%
Iran
3.4%
Britain/USA
2.0%
USA/Italy
3.0%
Serbia
2.0%
Italy
1.0%
Chile
7.7%
Mixed
2.0%
India
5.8%

USA
63.4%

Participants' fraction of People of Color

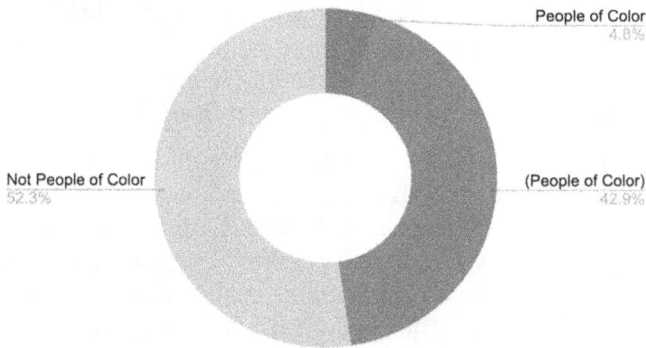

People of Color
4.8%

Not People of Color
52.3%

(People of Color)
42.9%

Figure 7.1. (Top) Participants seniority (%): junior (red), mid-career (yellow), senior (blue), (Middle) Participants nationality(%), (Bottom) Fraction of participants who were People of Color: People of Color (blue and red), Not People of Color (yellow). When in parenthesis, the fraction represents an assumed identity by the organizers based on personal knowledge of the participants.

During this session, participants were encouraged to reflect on their personal reasons to engage in anti-racism work within the Rubin Ecosystem, and discuss the definition of racism as a system of oppression and advantage based on race. Emphasis was placed on the distinction between diversity, inclusion, and equity and equality (and their relation to justice), as well as on the importance of performing a race-explicit analysis that names White Supremacy and anti-Black Racism. The second session was more interactive in nature, relying heavily on Zoom breakout rooms in order to overcome the limitations from the virtual format of the workshop. Participants were prompted by the facilitators to identify organizational and individual barriers to building racially equitable and pro-Black futures in the Rubin community, as well as promising practices to address racial inequities and advance opportunities for Black scientists in the Rubin Ecosystem. At the end, participants produced a document with concrete ideas to address anti-Black racism in their institutions and spheres of influence.[16]

7.4 Early Outcomes

Increased awareness of EDI and particularly awareness of systemic disadvantages faced by Black people in the USA and further afield has brought about a movement within Rubin Observatory Ecosystem to take immediate action. The workshop and the realization of the issues that it brought forth helped put a spotlight on systemic racism, anti-Black racism and their relation to EDI. This has resulted in most hiring managers requiring a statement from job applicants about their personal EDI efforts. These are a part of the evaluation process and are considered as important as technical/other skills. This is reinforcing the message that the value and importance of diverse teams is understood and internalized.

In addition, several months after the workshop we organized a follow-up session, via Zoom, in which we invited the participants of the workshop to (1) share how they have used the tools from the workshop, (2) share what actions they have taken inspired by the workshop, (3) collect feedback from the workshop, and (4) propose new ideas for next actions and future workshops. 17 of the about 70 workshop participants attended this follow-up session. In what follows, we reproduce particular examples shared by the participants of this follow-up session:

- o Construction Project Staff: I found what grabbed my attention from the workshop is the message that if you are not actively trying to seek and dismantle barriers, you are passively upholding them—I try to work within that framework in NOIRLab advocacy—bringing into the light the shared responsibility for increased representation. For example, when the organization published communications about Black History Month some of us found that a lot of the responsibility was put on people of color in our organization to produce these messages. This should not be the case. In addition to communications, we need to push the shared responsibility of addressing barriers in recruitment when serving on hiring committees.

---

[16] Document: https://padlet.com/ELteam/vibxvs1uff440un2

o University Faculty and Science Collaboration Lead: Faculty recruitment took on much of the advice that were raised in the last workshop, including wording and advertising to other audiences. We encountered positive reactions from the University but without directable action to enable them, so we came up with our own strategy to ensure diverse and unbiased recruitment for these positions. There were also difficulties establishing positions that are explicitly designated for individuals holding specific identities due to legal issues. For the faculty search, we had an explicit item in rubric for contribution to EDI. The identity of the candidate can be a direct contribution to the diversity, equity, and inclusion of the department.

o University Faculty and Science Collaboration Member: USCS insists on protocols and the academic personnel office runs training relating to how to review the EDI statements. Yet at UCSC we do not have a Black population at all. We have added a focus group of our diversity and climate committee on Black representation cause based on our population breakdown we just often miss a Black representation. Over the past year we created a safe space in our division specifically aimed at making space for Black students inspired by the American Institute of Physics National Task Force to Elevate African American Representation in Undergraduate Physics & Astronomy (TEAM-UP) report,[17] we are taking each of the pieces recommended by the TEAM-UP report and also representing Rubin at National Society of Black Physicists (NSBP) meetings.[18]

o LSSTC Science Staff: The LSST Corporation has run a selection process for postdoc fellows (Catalyst Fellowship), and I leveraged what I learned in the workshop as I was overseeing the selection process. Some info from the workshop helped us pass some of the new practices that we were concerned might get pushed back. The reviewers actually indicated that the anonymization of the process triggered a different kind of perspective. In summary what I learned was really to include EDI considerations in every step of the process, not as an add-on. In the LSSTC fellowship process, it was possible to craft the fellowship in a way that would be more attractive to Black candidates. [We] had the flexibility to create a program and emphasize career support, mentorship provided, and leadership training that would be provided, recognizing multiple ways of benefiting the community. We were able to create a position that would be more distinctive. We also had an application in two parts. Applicants write both a community impact statement (could be personal experience) and a research statement. These are weighted equally. Did not need to explicitly include diversity criteria to achieve a diverse pool of finalists because community impact was built into the process.

o University Faculty, Rubin Operations and Science Collaboration Member: After the workshop I took materials from the workshop and dedicated one day in the class in the semester to go over EDI topics. The feedback I got was

---

[17] https://www.aip.org/diversity-initiatives/team-up-task-force
[18] https://nsbp.org/page/2022NSBPConferenceSummary

that they [the students] never have physics faculty members talk about these things or have anyone tell them that being ognizant of these issues is an important part of their professional development and I was struck by how undergrads are not exposed to these topics at all. In the Physics department, I was able to make funds available for an undergraduate physics fellowship for students from underrepresented groups (could not be done for Black students because of legal issues). As I was researching how to set this up, I found out that Hubert Mack Thaxton, who I consider to be a significant historical figure, did his PhD thesis at our University on proton–proton scattering and we named the fellowship after him. I engaged into researching him to include information about him in the fellowship and his family reached out to me to thank us for this.

o Rubin Leadership Team: Rubin is committed and we have been working to implement some of the things we learned from this workshop, things similar to what our colleagues from the LSSTC referred to: trying to be aware of biases and take concrete steps in recruitment to avoid them. There are times, however, when you, if you are *not* biased, you are not making a difference. You need to be deliberate to include for example Black scientists in job searches or even think of targeted hires as many organizations do for many reasons. We have not been successful at hiring a Black scientist. In the current searches, we have no Black scientists in the applicant pool.

o Rubin Construction/Operations Staff: Another thing we tried is Land Acknowledgement. We tried to include it for a few presentations. It was considered not completely appropriate with its wording. Continuing to work on this. Indigenous members of the NOIRLab community are helping with the message.

o The participants' experiences can be grouped into several themes. One of these themes was the importance of actively seeking and breaking down barriers, and that the responsibility for increasing representation should not rest solely on individuals from underrepresented groups. Another theme was the need to include EDI considerations in every step of the process, rather than as an add-on. Participants reported success in creating fellowships and positions that were more attractive to Black candidates by emphasizing career support, mentorship, and leadership training, and by creating an application process that weighted community impact equally with research statements. They also described creating safe spaces for Black students, and developing class materials to educate students about EDI topics. Legal issues were noted as barriers to some initiatives, such as creating positions explicitly for individuals with specific identities or funding fellowships for Black students. Some participants reported that despite their efforts, they had not been successful in hiring Black scientists, and suggested that targeted hires might be necessary. Overall, the participants found the workshop to be a valuable tool for advancing EDI efforts in their respective fields.

7.5 Lessons Learned and Challenges

- *Be clear about the topic:* Rubin is an international collaboration and equity does look different in different local contexts. While anti-Black racism is demonstrably a global issue, different cultures experience the specificity and criticality of anti-Black racism differently. This resulted in a range of reactions that included participants asking why we were being so specific addressing only anti-Black issues and participants feeling like their specific experiences of marginalization had been overlooked. Nonetheless, the organizers stand by the choice of focusing on a narrow and timely issue which allowed the conversation to be specific and detailed. The organizers' own experience with more general "anti-discrimination" workshops, or even generalized anti-racism work is that broad topics end up in forcing generalization that do not allow for specific planning on effective action. After the first session, one of the participants expressed their frustration about how certain participants kept centering themselves instead of Black people and scientists in the conversation. Ultimately, whatever topic is selected by organizers, we suggest specificity.

- *Ensure commitment to follow-up after the workshop and report back*: It's normal that momentum is lost almost immediately after the workshop is over, so before the first workshop, it is useful to set a check-in schedule, and ask people to commit to sharing their EDI work and perceived outcomes (negative and positive) so that everyone can learn from them. In our case, we initially considered gathering feedback from the workshop participants via a Google form after the workshop. In the end, as indicated above, we organized a follow-up session, via Zoom, in which we invited the participants of the workshop to (1) share how they have used the tools from the workshop, (2) share what actions they have taken inspired by the workshop, (3) collect feedback from the workshop, and (4) propose new ideas for next actions and future workshops.

- *Collect data to measure impact*: It can be difficult to measure success. Whether someone feels a sense of community or belonging are quite intangible. Other metrics can be quite concrete, such as measuring over time if the number of Black job applicants has changed, keeping track of the number of subgroups or ecosystem members involved in proactive EDI activities, and recording the number and kinds of EDI-related activities that result after the event to compare with those before. However, attributing such a change back to actions taken by the organization or its individuals is challenging. In addition, it is also challenging that some organizations do not allow such data to be collected and even if it is, Human Resources departments are reluctant to share it. Insight can also be gained by hiring social scientists or other professionals who can inform on how to gather these metrics properly, perhaps via hiring organizations to conduct climate surveys or other mechanisms, and who can provide suggestions on how to act based on the findings.

- It was not clear if involving Black and African American Rubin members in the workshop resulted in any benefit. Some of the Black and African American participants reported feeling that the workshop was not effective for them and instead it put them in the position of having to hold participants' hands through realization of a reality that is all too obvious to them. Conversely, we wonder if a workshop on anti-Black racism might have felt more abstract and academic if the participants did not have the Black members of this community under their eyes at the time of the workshop. If minority colleagues are included or asked to be included, monetary compensation or compensation in the form of enhanced job-security should be offered.[19] In addition, analogously to "affinity spaces", "solidarity spaces" could be created for white people only to educate themselves and ask learning questions without burdening their minority colleagues.[20]

Acknowledgements

We thank Dara Norman, Dorian Russell, and Tim Sacco for the invitation to write this contribution and for editorial and content suggestions that improved the manuscript, Brian Nord for important feedback and discussions during the writing of the LSSTC Enabling Science 2020 proposal that partially funded the workshop, the LSST Corporation and the leadership of the Rubin Observatory for funding the workshop, Las Cumbres Observatory for administration of the LSSTC grant, the workshop and follow-up session participants, *The BIPOC Project*, Fiona Kanagasingam and Merle Mcgee for facilitating the workshop, the organizers of the virtual Rubin Project and Community Workshop 2021, Anissa Tanweer for feedback on the manuscript, and the additional co-PIs of the LSSTC proposal for their support (Amanda Bauer, Jeffrey D. Barr, William Brandt, Patricia Burchat, Rachel Mandelbaum, Brian Nord, Chad Schafer, Sandrine Thomas). The work of AAPM was supported by the U.S. Department of Energy under contract number DE-AC02-76SF00515.

---

[19] Suggestion by B. Nord (Fermilab), in the context of the Dark Energy Survey project.
[20] Suggestion by T. Jackson during their plenary talk in the virtual AIP TEAM-UP Implementation Workshop of January 2021, https://www.aip.org/diversity-initiatives/teamup-implementation-workshop-1

Part III

Changing Culture and Fostering Inclusion in Collaboration

Introduction to Part III: Changing Culture and Fostering Inclusion in Collaboration

The goal of Part III is to showcase efforts to make inclusive cultural change in favor of more equity in astronomical collaboration and institutions. Policymakers and scholars alike have lauded collaboration for its ability to produce high-quality research, answer complex questions, and generate "more nuanced and robust understandings" of our Universe (Frickel et al. 2016). Funding organizations have increasingly supported collaborative research efforts as well, as it is associated with heightened creativity among team members or a broader reach of their scientific outputs. When successful, research collaboration has the capacity to push the boundaries of knowledge and craft new lenses through which we understand our place in the Universe.

In theory, collaboration is so beneficial because it brings together team members with diverse expertise and viewpoints, thus facilitating new approaches for thinking about complex problems. However, Michelle Bennet—in one of the contributions to this section—highlights what she calls "the ultimate paradox of Team Science": that difference between team members is both the driver of novelty of a project, while those same differences can drive tensions and undermine inclusion on the team. Even collaborative teams with representational diversity among its members can employ practices that create barriers to full inclusion on the team. These collaboration-level dynamics that cause inequalities are further exacerbated by organizational processes that reinforce barriers for historically underrepresented groups.

From this foundation, the contributions to this section highlights various efforts to disrupt the processes that drive inequalities in astronomical collaborations and institutions. First, *Team Science: An Exercise in Difference and Diversity* puts forth best practices for creating an inclusive team environment, and stresses the importance of team policies, processes, and procedures in facilitating inclusive teams. The second contribution, *"How Grantmaking Organizations can further DEI goals"* highlights some early efforts of funding agencies to drive more inclusive collaborations in astronomy through the mandate of inclusion plans. The contribution argues that the organizations with control over research funding or telescope time are uniquely poised to encourage equitable collaborations by tying DEI gains to the allocation of resources.

The final contribution, *"Building and Maintaining a Culture of Inclusion in Astronomy and Astrophysics"* also argues for improvements to the culture of astronomy in order to be more inclusive of underrepresented groups. The contribution highlights how current cultural norms in astronomy undermine inclusivity in the field, as does the unequal access to resources and the devaluation of activities like software development and public engagement. The piece advocates for redefining scientific merit to include DEIA principles. Taken together, these contributions highlight structural changes that can be employed at varying levels throughout astronomy and astrophysics in order to facilitate a more inclusive culture and broaden participation from historically marginalized groups.

AAS | IOP Astronomy

An Astronomical Inclusion Revolution
Advancing Diversity, Equity, and Inclusion in Professional Astronomy and Astrophysics
Dara Norman, Tim Sacco and Dorian Russell

Chapter 8

Team Science: An Exercise in Difference and Diversity

L Michelle Bennett

My expertise is in Team Science, which I often describe and view as an exercise in difference and diversity. Team Science values, at its most fundamental level, difference, and requires the team to create an environment of psychological safety. So, as I work my way through this piece, I will be describing what I know best and using that as my foundation for making the case that it is valid to propose that valuing difference in team science, at its core, can increase diversity, inclusion and belonging. I suggest that is because difference and diversity are required for teams to innovate and solve the challenging scientific problems before us through discovery, inquiry, and innovation. Teams can do this maximally when they feel safe to share, question, take risks, and propose new ideas. The feeling like you can put forward an idea and know you will not be laughed at, retaliated against, or undermined is what it is to have psychological safety (Edmondson 1999, 2018).

My work over the last couple of decades in Team Science and Collaboration has made me particularly sensitive to the notion that, in order to solve the most complex scientific questions facing society today, we must bring people together from different disciplinary backgrounds to establish a shared theoretical framework that integrates, melds and synthesizes views, approaches, and ideas (Rosenfield 1992). The people who come together and strive to solve perplexing questions must have the freedom and time to innovate, which requires working in an environment that both encourages it and supports it.

What drives innovation? Innovation occurs in the gap or emptiness between two knowns. This means finding, identifying, and making new connections between things that were previously not linked (Dyer et al. 2019). Those new connections and newly found associations serve as the foundation for developing new tools, creating technologies, or designing novel approaches. It is often argued that if a problem were simple, it would have already been solved.

doi:10.1088/2514-3433/ad2174ch8　　　　8-1

To make new connections, one must feel comfortable being creative, tossing out wild ideas, and then batting them around and building on them. It is essential to be open minded about the multitude of possibilities and potential for linkages not initially observed. Here is where difference comes in. Each dimension of differences a team member brings "to the table" contributes to an increasingly complex way of looking at a problem and making connections between and among things that, at first glance, do not have anything to do with each other. Herein lies both the opportunity team science presents as well as its challenges. If people feel as though they are in competition as opposed to in collaboration, working through the differences can be long and messy, as opposed to fun and energizing.

This leads to describing the ultimate paradox of Team Science. The complex challenge before the team is that it requires different disciplinary approaches, expertise, and methodological approaches, among other things, to succeed. It sounds easy, doesn't it? Let's get a lot of people in a room together who can all think differently about a problem, and we'll knock it out. It is reasonably safe to assume that the people who come together are different with respect to how they think, what they perceive, and the level of risk they are willing to take. They will come from diverse backgrounds, have had a spectrum of lived experiences, and distinctive personalities, communication, conflict, and management styles. And of course, they will have been trained in different disciplinary cultures, each with their own language, set of values, and norms. These differences that are essential for innovation can be maximized by creating a safe environment, by establishing trust and psychological safety as the team is built.

8.1 Creating a Team Dynamic

To create the conditions desired for Team Science and for working well together, it is not unusual to recognize that behaviors need to change. For example, co-creating a vision, or jointly deciding on roles and responsibilities, or sharing credit, data and preliminary data may require exercising some new skills. In this way individuals must examine and alter their behaviors or develop new abilities so that they align with how the team wants to operate. Many leadership or team training programs, modules, and workshops focus on the behaviors needed to work together well, such as negotiating skills, having difficult conversations, setting expectations, establishing a vision, and of course establishing trust and psychological safety. This includes my own work in Team Science and collaboration which initially focused largely on the behaviors the team and its members needed to be successful (Bennett et al. 2018).

What I was missing, and what I believe many are missing now, is that to do its most impactful collaborative work at the team or at the organizational level, more than just behavior change is required. Team behavior and norms derive directly from the team mindset, or culture. When individual or team mindsets are in alignment with their behaviors, then the behaviors can be consistently and reliably reproduced. However, if the mindset is at odds with the desired behaviors, it is very difficult to continue behaving in a way that aligns with those mindset elements. This is especially true when stress levels increase such as with conflict or an unexpected urgent situation. In stressful situations, the comfortable, well established habitual behaviors emerge and if they align with a unilateral control mindset, it will lead to more difficulty in working successfully with others (Schwarz 1994, 2013).

The unilateral control mindset and the mutual learning mindset have been written about extensively by Roger Schwarz (Schwarz 1994, 2013) who adapted them from the work of Argyris and Schön's Model 1 and Model 2 (Argyris & Schön 1974, 1978; Argyris 1982). Model 1, or the unilateral control mindset, is a power-oriented approach used to control a situation when in conversation with others. Most of us can access both models and are socialized to become experts in unilateral control by an early age, based largely on what we are exposed to as we are raised. Model 2 or the mutual learning approach focuses on understanding that people see things differently and by working together all involved can learn something new. Inherent in the approaches are both mindset (values and assumptions) and behavior elements.

I have come to believe that when the mutual learning mindset serves as the foundational mindset for a team, a department, a unit, or an organization, it enables productive conversation about anything that is important, ranging from the highly technical to a focus on relationship elements. In both situations mutual learning lays the foundation for highly productive conversations even about difficult, emotional, conflict-generating topics. Among the very difficult conversations before us as a society are those related to diversity, inclusion, and equity. This is what leads me to focus so heavily on mindset in the context of Team Science and by extrapolation, to the intersection of team science with diversity, equity, and inclusion (DEI) since I believe that diversity and difference enable Team Science. When teams are very safe places for their members to ask questions and be creative in, diversity and difference are fully leveraged. Psychologically safe environments highly permissive of questioning and creativity successfully leverage all the diversity and differences.

8.2 Team Effectiveness Model for Science

In its most simplified form, innovation requires difference, and differences can lead to disagreements. So, creating a safe environment where differences can be expressed, queried, understood, and leveraged helps the team. In essence, the differences that are so essential can form the foundation upon which teams do their best work. Science team results can include conducting highly innovative research, supporting each other's career growth and development, and experiencing a sense of belonging as a group member. The difference and diversity that serves as the foundation for the results could include disciplinary, methodological, world view, personality, lived experience, philosophy, and perspectives, as well as others.

I recently co-authored a piece introducing the Team Effectiveness Model for Science (TEMS) (Schwarz & Bennett 2021). The TEMS can serve as a road map for newly forming teams and as a diagnostic and intervention tool for existing teams. It has at its foundation the adoption of the mutual learning mindset and behaviors. From here the team is encouraged to identify values and behaviors for conducting their scientific work as well as values and behaviors for how they will be in relationship with each other. This is in fact an approach for developing team culture. Culture refers to the shared values and beliefs of the team or organization, which are then continually reinforced by the leadership and subsequently by all members of the team or organization. How people behave in the organization is a visual expression of the values and beliefs (i.e., the culture) of the organization (Chatman & Eunyoung 2003; Kerr & Slocum Jr 2005).

Building from there, the next step of the model is for teams to purposefully establish their structures and processes for their science and for their relationships. The notion that a team can develop these on their own is extremely empowering for the team. The team does not have to leave anything to chance. The members can design the structures and processes in a way that fully aligns with the values and behaviors they have identified and agreed will ground them and their work together. For example, the vision and mission, roles and responsibilities, and membership are all developed and agreed to by the team. Processes such as how the team will make decisions, how they will run meetings, how they will hold each other accountable, and what they will do when conflict erupts can all be designed by the team. As the structures and processes are put in place, the team will want to ensure that their designs are based on the values they have established, otherwise the structures and processes risk being incongruent with the values and risk not supporting achieving the results the team desires.

What do you do when a team has already been formed and structures and processes are already in place? This is when you ask the question as to whether the structures and processes align with the mindset the team would like to be operating from. For example, if the team says equity is one of its values, then one could examine the team policies, structures, processes, and procedures and assess whether they inadvertently contribute to inequity.

This is highly relevant when we consider another challenge many teams and organizations are facing which is focused on a goal of dismantling structural racism. If a team or organizations says it wants to dismantle structural racism, it must then assure that the structures, policies, and processes in place do not inadvertently work against that desire. Making this assessment is key because when identified, it can be purposefully redesigned to align with the desired organizational mindset, or culture, that is free from barriers to equity. Joshi and Jackson provide an example in their writing of a Human Resource system that limits the promotion of women and minorities to higher level positions within an organization due to the existence of glass ceilings and walls (Joshi & Jackson 2003). In an organization like this, it is unlikely that the development of diverse teams will result in greater productivity or hoped for performance improvement given the lack of upward mobility to some team participants. Encouraging diversity at the entry level of organizations but not supporting the pathway for career development to the top creates a structural barrier. The barrier to upward mobility is an example of an existing structure and process that would need to be redesigned to align with the stated values of the organization.

In considering how a group might go about looking at the policies, processes, and structures of an organization, I believe it is essential that an environment of psychological safety exist to have those conversations and achieve successful outcomes.

8.3 Diversity in Team Science

Science teams may have very different scientific results they want to achieve and yet it is safe to say that teams also have some fundamental results in mind, many of which align with what it means to be successful. Among them are increased innovation, more effective problem solving, and publishing success. Just as important as the scientific results are those for team relationship and individual well-being.

What is it about diversity and difference that catalyzes innovation? If one accepts the notion that innovation occurs in the gap between two knowns, the question then becomes how do we find or identify a thread to connect them. It is essential to be open to the processes of exploration and discovery as a way to begin identifying possible ways in which one thing might relate to another. In addition, curiosity, the art of asking genuine questions, those that provoke thought, reflection, and insight enables the building of theoretical frameworks from what is discovered to establish connections. If a team member does not feel safe exploring, asking questions, sharing ideas or thoughts, they will not. If one person will not, it may be that others do not either. Then the team is at a huge disadvantage because the talents, strengths, and ideas of individuals invited to the table are, well, never put out there for others to build on or co-create with.

Experiments performed almost two decades ago revealed that when you bring a group of experts together to form a team to solve complex problems, they underperform as compared to a team of highly intelligent individuals randomly chosen to work together. The reason being that the more diversity the team had, the more innovative it was at solving the assigned problems (Hong & Page 2004). The authors observed that the more heterogeneity a team had, the more challenging the social dynamics. This might help explain why people tend to feel more comfortable collaborating when they see themselves as being alike (Gratton & Erickson 2007).

In another example, research published by Joshi and Jackson focused on workplace diversity in organizations led a number of interesting findings, including the fact that diversity within a team can cause some social tensions (Joshi & Jackson 2003). They added a note that it is important to be mindful of this at the organizational level as teams are put together. At the same time, that team diversity led to an ability of those teams to maximize their connectivity with other teams in the organization to a greater extent than more homogeneous teams.

Woolley and Malone performed research to understand the role of collective intelligence in teams. Interestingly the IQ of different team members did not correlate with the measure of collective intelligence and neither did some well-used measures of group interaction such as satisfaction and cohesion. The factor that seemed to play the greatest role in raising the collective intelligence of the group was the presence of women on the team (Woolley & Malone 2010). This research was used, in part, to make a case for the California Law SB826 signed into law in 2018 requiring publicly traded companies to include women on their boards by the end of the following year.[1]

Enactment of this law resulted in an increase in the number of women serving on boards and it forced individuals who asserted there was a paucity of qualified women to serve on boards to recognize that as nothing more than a myth. As a result of adding women, the cultures of the boards started changing, especially changes related to governance. Companies that brought on women, typically from outside of the usual network of colleagues, observed a variety of benefits including smarter mergers and acquisitions, more innovation, and improved handling of crisis situations (Covert 2022).

8.4 Aligning Policies, Processes, and Procedures with Team and Organizational Values

This leads me to the importance of the organization and the implementation of policies, processes, and procedures that are in alignment with its values. If the organization states that it values team science or DEI, it should be evident by examining the way that the organization functions and how it treats its people. An area of great frustration in the realm of team science is related to recognition and reward. I have found this to be especially true in the biomedical sciences where the individual often seems valued over a scientific team. As a result, when promotion and tenure policies do not support team science, it is very difficult to create a path for the team scientists into more senior positions within the organization where they could have more influence on the same.

Similarly, if existing policies, processes and/or procedures create barriers for members of underrepresented groups to advance in an organization, the recruitment of these individuals into the organization will not necessarily achieve the stated goals of the organization. If, on the other hand, organizations look carefully at how they retain, promote, recognize, and reward individuals and teams, and ensure that the approaches are congruent with the values they say they hold, then there will be progress in assuring that the composition of the whole organization will reflect the values the organization holds for supporting and advancing its people. The diversity —in all dimensions—can serve as an asset by enabling innovation (Dyer et al. 2019). Scientists, research teams, and the institutions they work in can create and foster equitable workplace environments when they adopt a mutual learning mindset.

References

Argyris, C. 1982, Reasoning, Learning, and Action: Individual and Organizational (Hoboken, NJ: Jossey-Bass) 1st ed.

Argyris, C., & Schön, D. A. 1974, Theory in Practice: Increasing Professional Effectiveness (Hoboken, NJ: Jossey-Bass) 1st ed.

Argyris, C., & Schön, D. A. 1978, Organizational Learning: A Theory of Action Perspective, (Reading, MA: Addison-Wesley)

---

[1] This law is currently (2022) being challenged in court.

Bennett, L. M., Marchand, C., & Gadlin, H. 2018, Collaboration and Team Science: A Field Guide, https://www.lmbennettconsulting.com/resources

Chatman, J. A., & Eunyoung Cha, S. 2003, Calif. Manage. Rev., 45, 19

Covert, B. 2022, Politico Magazine, https://www.politico.com/news/magazine/2022/02/25/california-companies-women-boards-quotas-00010745

Dyer, J., Gregersen, H., & Christensen, C. M. 2019, The Innovator's DNA (Brighton, MA: Harvard Business Review Press) Revised edition

Edmondson, A. C. 1999, Adm. Sci. Q., 44, 350

Edmondson, A. C. 2018, The Fearless Organization (New York: John Wiley & Sons) pp 3–24

Frickel, S, Albert, M, & Prainsaclk, B 2016, Investigating Interdisciplinary Collaboration: Theory and Practice Across Disciplines (New Brunswick, NJ: Rutgers Univ. Press)

Gratton, L., & Erickson, T. J. 2007, HarBRv, https://hbr.org/2007/11/eight-ways-to-build-collaborative-teams

Hong and Page 2004, PNAS, 101, 16385

Joshi, A., & Jackson, S. E. 2003, *In:* International Handbook of Organizational Teamwork and Cooperative Working, ed. M. A. West, D. Tjosvold, & K. G. Smith (West Sussex: Wiley)

Kerr, J., & Slocum, J. W. 2005, Acad. Manage. Exec., 19, 130

Schwarz, R. M. 1994, The Skilled Facilitator: Practical Wisdom for Developing Effective Groups (Hoboken, NJ: Jossey-Bass)

Schwarz, R. 2013, Smart Leaders Smarter Teams: How You and Your Team Get Unstuck to Get Results (Hoboken, NJ: Jossey-Bass)

Schwarz, R. M., & Bennett, L. M. 2021, J. Clin. Transl. Sci., 5, e157

Rosenfield, P. L. 1992, Soc. Sci. Med., 35, 1343

Woolley, A., & Malone, T. 2010, Sci., 330, 686–8

AAS | IOP Astronomy

An Astronomical Inclusion Revolution
Advancing Diversity, Equity, and Inclusion in Professional Astronomy and Astrophysics
Dara Norman, Tim Sacco and Dorian Russell

Chapter 9

How Grantmaking Organizations Can Further DEI Goals

Timothy Sacco

9.1 Introduction

Every ten years, the National Academies of Science, Engineering, and Medicine facilitates a survey of the Astronomy and Astrophysics (hereafter Astronomy) field to determine what areas of science should be prioritized in the coming decade. These "decadal surveys" are consensus documents produced by the community, which federal agencies, like the National Science Foundation (NSF), NASA and Department of Energy draw on when crafting future Astronomy policy. The most recent survey, "Astro2020" (NASEM 2021), explicitly advocated for prioritizing diversity, equity, and inclusion (DEI) in Astronomy. While Astronomy has become more diverse over time, women, people of color, and other marginalized groups remain underrepresented among earned Astronomy PhDs (Porter & Ivie 2019; Prescod-Weinstein 2021). Sociological studies on scientific research demonstrate that both organizational forms (Acker 1990; Britton & Logan 2008; Wooten & Couloute 2017; Mickey 2019; Smith-Doerr et al. 2019; Ray 2019) and collaborative dynamics (Smith-Doerr et al. 2016; Jimenez et al. 2019; Barber et al. 2020; Eaton et al. 2020; Stevens et al. 2021; Jeske et al. 2022) reproduce inequalities in science. Astro2020 argues that a more concerted effort must be made to the organization of astronomical research—such as how collaborations are structured to do science—in order to make the field more equitable. Astro2020 puts forth several recommendations for science, technology, engineering, and mathematics (STEM) policymakers to craft mandates that incentivize the integration of underrepresented groups in Astronomy through the organization of research collaborations.

Today, across scientific fields, far more scientific research is collaborative than sole-authored (Bozeman et al. 2012; Leahey 2016). Policymakers have consistently supported research collaboration for its ability to produce high-quality research, answer complex questions, and generate "more nuanced and robust understandings" of our Universe (Frickel et al. 2016). Funding organizations have increasingly supported collaborative research efforts as well (Bikard et al. 2015), as it is associated with heightened creativity among team members or a broader reach of their scientific outputs. When successful, research collaboration has the capacity to push the boundaries of knowledge and craft new lenses through which we understand our place in the Universe (Wuchty et al. 2007). However, as I will review below, the social processes around research collaboration and the organization of scientific research production more broadly has been shown to reproduce social inequalities. How can policies target astronomical collaborations in a way that further DEI goals in the field?

In this contribution, I argue that organizations that are in control of the resources necessary to do astronomy—like research funding or telescope time—are uniquely poised to mandate policies that can change on-the-ground behavior in favor of DEI goals. I outline how the social dynamics of both research collaborations and larger scientific organizations have reproduced inequalities in astronomy and astrophysics. Then, I highlight a growing social science scholarship that investigates how "grantmaking organizations," (GMOs)—those in control of the resources necessary for research—can further DEI goals by incentivising those goals through access to resources. I describe the social science scholarship on how GMOs can engage in "equity change work" to drive DEI goals in science. Then, I detail two cases of GMOs engaging in equity change work in astronomy and astrophysics by connecting research resources to the furtherance of DEI goals.

9.2 Background

In recent decades, potential scientific workers in the US are increasingly diverse, yet inequalities persist in both astronomy and astrophysics, and in STEM more broadly. Within Astronomy, men make up roughly two-thirds of the PhDs awarded annually (Mulvey & Pold 2023). Since 2000, there have been around 100 Astronomy PhDs granted annually. According to the most recent census, African Americans make up *12.6%* of the population, but there have been only *one or two* Black PhD earners in astronomy per year over the last decade. Similarly, Hispanic or Latinos make up *18.9%* of the population, but while the numbers have risen in recent years, they still are a disproportionately low fraction of PhD earners compared to their overall representation in the US population.

Inequalities among astronomers and astrophysicists are often exacerbated by the social dynamics of research collaborations and the organization of the astronomy field. Scholarship that explores diversity in collaboration has repeatedly found evidence that team members from underrepresented groups are assigned devalued but necessary work in collaborations at significantly higher rates than their majority counterparts (Jimenez et al. 2019; Barber et al. 2020; Eaton et al. 2020; Thomas et al. 2020; Stevens et al. 2021). These devalued tasks include practices like education, mentorship or outreach. One study on the collaborative experiences of biomedical researchers found that underrepresented team members were often treated as experts on diversity first and foremost, which led to the minority scientists to be reassigned outreach tasks on the project (Jeske et al. 2022). As a result, underrepresented team members saw limited opportunities for upward mobility or to substantively contribute their expertise to their teams. Collaborative dynamics can also facilitate inequality around resources, credit, and visibility among team members (Smith-Doerr et al. 2016).

Within science, the organization of funding agencies and the systems that determine the allocation of resources has been found to reinforce inequalities among collaborators (Hackett 1990; Walsh & Lee 2015; Lee & Walsh 2022). The rules and regulations around funding, telescope time, or tenure have been shown to exacerbate inequalities in both astronomy and in STEM fields more broadly (Acker 1990; Britton & Logan 2008; Wooten & Couloute 2017; Mickey 2019; Ray 2019; Smith-Doerr et al. 2019). For instance, federal funding calls may appear to be neutral to all applicants, but applicants at institutions with more sophisticated infrastructure for doing science or support for grant-writing maintain a competitive advantage when submitting a research proposal. One study shows that the institutions with more advanced scientific resources and administrators who identify and oversee grant submissions tend to be less diverse than smaller institutions without these resources (Kameny et al. 2014). This leaves scientists at smaller institutions, who are more likely to be underrepresented in STEM than their research university counterparts, at a structural disadvantage. The dynamics of collaborations may reproduce inequalities, but collaborations are in part a byproduct of the organizational environment in which they are embedded.

Ultimately, this system creates what sociologist Robert Merton (1968) referred to as a "Matthew Effect of accumulated advantage." Merton developed the concept from the bible verse Matthew 25:29, which reads "For everyone who has more will be given, and [they] will have abundance; but from [they] who [have] not, even what [they] have will be taken away." In other words, the rich get richer and privilege begets privilege. The organizational processes that guide research collaborations tend to reproduce privileges for those who are already privileged in scientific work. Simultaneously, those astronomers at smaller or under-resourced institutions that have been historically underrepresented in the research process remain so due to their lack of access to the resources needed to participate, different administrative demands around teaching or service, and a dearth of ties to more well-resourced astronomy institutions and professionals. Changing this landscape will require intentional action by those in a position to do something.

9.2.1 How GMOs Can Drive Change

For years, institutional scholars have investigated the ways that organizational processes reproduce inequalities in science. Recently, a subset of this community has theorized how organizations can implement policies and practices that will further DEI goals. McCambly & Colyvas (2022) use the phrase *racialized change work* to signify intentional organizational strategies that are designed or mandated to further DEI goals. Here, I broaden the term to *equity change work*, not to detract from the necessary efforts to further racial equity in astronomy and beyond, but rather to capture the ways that organizations are engaging in this work. Astronomy organizations have begun implementing policies that, broadly speaking, broaden participation in ways not limited to activities focused on racial equity. While some policies or programs target underrepresented individuals in STEM, while others target underrepresented institutions. All these policies are focused on broadening participation in STEM, albeit in different ways. Thus, inclusion change is meant to characterize this broad movement that is unfolding within the scientific field.

Within STEM contexts, GMOs are in a unique position to engage in equity change work, especially through policies that target the research process. First, GMOs can promote DEI goals by leveraging control over valued resources, like research funding or telescope time. Federal GMOs also tend to have advanced policy agendas, and are poised to leverage their control over resources as a way to implement ideas into policy (McCambly and Colyvas 2022). The argument that GMOs are in a position to implement equity change work to drive DEI goals through control over resources is strengthened by qualitative social science research that demonstrates how scientists tailor their research programs to the interests and requirements of their funders (Kleinman 2003; Jeon 2019; Reinecke 2021). Taken as a whole, GMOs can be an important institutional source for change towards DEI in astronomy and other STEM fields through attaching restrictions or requirements to the resources needed to support astronomy and astrophysics research.

Thankfully, GMOs have already begun implementing policies that tie research support to DEI goals. Hunt et al. (2022) reviewed the policies of 23 GMOs from 18 different countries. Their review is written with funders in mind; by providing an overview of policies that have been implemented, the paper provides a framework for funders to continue engaging in equity change work. Based on their review of preexisting policies, Hunt et al. put forth best practices for how GMOs can effectively engage in equity change work by mandating DEI practices as a stipulation for resources. First, they encourage agencies to come up with clear definitions for important concepts that should then be shared with applicants. Sharing concepts with applicants before they submit proposals sets them up to structure inclusive research collaborations. Next, GMOs must determine when and how to mandate DEI policies, including decision making around when DEI will be suggested versus mandatory, and what types of instructions will be provided to proposers. GMOs also tend to see evaluators as important for assessing the successfulness of policies. The authors urge GMOs to develop clear and concise instructions for evaluators to ensure projects are meeting the mandated goals. Hunt and Scheibinger also encouraged GMOs to evaluate their own DEI policy implementation processes in order to improve their influence on inclusion in scientific research.

Many applicants, evaluators, and administrative staff will need training in order for mandated policies to yield their desired goals. Hunt et al. argue that these trainings should also be the responsibility of funding agencies. If not for GMO training, applicants and others on the project would likely access training through their host institution, likely a college or university. They note that these kinds of training are not consistent across universities. We know that higher education is fraught with resource inequalities, which could shape the types or frequency of training offered at a given institution. We also know that, in the current political climate, there is significant variation across state-level policies in regard to DEI. As a result, as GMOs require that researchers meet certain requirements to be eligible for research resources, applicants will benefit if the GMO offers training that outlines how researchers can navigate these mandates. Training should outline important concepts, knowledge, and expectations.

9.3 U.S. Astronomy & Astrophysics

This scholarship on the specific ways in which GMOs can drive DEI goals are a useful foundation for furthering DEI goals in astronomy and astrophysics. Within the U.S. astronomy and astrophysics community, funders take queues from the decadal surveys. DEI goals had been gaining traction throughout the 2010s, and so by the time the decadal survey began it was clear that workforce concerns including DEI needed to be included in the survey. Astronomers organized white papers targeting the decadal survey through professional meetings, including several Women in Astronomy meetings and two Inclusive Astronomy conferences; one in 2015 produced a comprehensive white paper on achieving inclusion in Astronomy (*Nashville Recommendations 2015*), and a second in 2019 inviting underrepresented Astronomers to draft white papers with recommendations for the 2020 decadal survey. Several of these white papers appeared together in a special issue of the Bulletin of the American Astronomical Society, which publishes reports and commentaries about Astronomy and Astrophysics.

The most recent decadal survey, Astro2020, argued that "the pursuit of scientific excellence [in astronomy] is inseparable from the humans who animate [that science]". Astro2020 highlighted several cases in which GMOs had supported DEI initiatives in astronomy and astrophysics education (NASEM 2021: 124–125). Yet, as detailed above, Astro2020 urged more gains in promoting DEI policy in astronomy and astrophysics research. Astro2020 highlights some successful cases of funders improving equity in research, such as "NASA's Hubble Space Telescope [being] the first to employ a dual anonymous proposal review process in 2018, after analysis of gender-based proposal successes over 10 years demonstrated a small but consistent pattern of male PI success disproportionately exceeding that of women's success" (p. 134). Astro2020 encourages GMOs in the field of astronomy and astrophysics to identify other organizational ways to foster greater equity in astronomy and astrophysics research.

9.3.1 Promoting Inclusive Collaborations through "Inclusion Plans"

Inclusion Plans are one empirical example of GMOs using control over resources as a way to promote more equity and inclusion in astronomy and astrophysics collaborations. As of the time of this writing, a few funding agencies or programs have implemented a requirement for inclusion plans alongside research and data management plans as part of the research proposals for funding. In an inclusion plan, proposers describe how the organization of their research collaboration furthers broadly defined DEI goals within astronomy and astrophysics. Highlighted here are two examples of implementations: NASA's Astrophysics Theory Program (ATP)—the first instance of inclusion plans to be mandated in the field—and the U.S. Extremely Large Telescope Program (US-ELTP), which will mandate inclusion plans in proposals for the awarding of telescope time, once the affiliated observatories are online.

9.3.1.1 NASA Astrophysics Theory Program (ATP)

NASA's ATP was the first instance of inclusion plans being mandated in astronomy and astrophysics. In 2020, NASA added "Inclusion" to its set of core mission values, complementing their existing values of Teamwork, Safety, Integrity, and Excellence. NASA's Astrophysics Division implemented ATP inclusion plans as a strategy to further DEI goals within the space sciences. ATP financially supports theoretical astrophysics research, the development of astrophysics models, and facilitates the interpretation of space astrophysics data. To be eligible for funding, proposals had to submit an inclusion plan alongside their science proposal.

Because NASA ATP was the first instance of inclusion plans being mandated in astronomy and astrophysics, there was little precedent for what inclusion plans should look like or how they should be assessed. As a result, ATP organized two panels with distinct expertise to review the inclusion plans. The panel of astronomers were tasked with reviewing all parts of the proposal, i.e., the scientific goals and the inclusion plans. The second panel included reviewers with DEI and STEM expertise, including some astronomers. Dr Dara Norman and I were able to work with NASA to obtain information about reviews and grades in order to provide an assessment of the experiment to NASA ATP leadership. A redacted version of the report has been published in the Bulletin for the American Astronomical Society (Sacco & Norman 2022) and detailed findings can be found there. Important here, however, is that this experiment took place and ATP used the report to formalize the assessment of inclusion plans for future reviews. They used the process and the report to identify blind spots, which then shaped the guidance and resources provided to the astronomical community for the future preparation of inclusion plans, including a workshop on good practice for writing inclusion plans.

9.3.1.2 NOIRLab's U.S. Extremely Large Telescope Program

A second example of inclusion plans in astronomy and astrophysics is one planned as part of the NSF's National Optical Infrared Research Laboratory's (NOIRLab's) U.S. Extremely Large Telescope Program (US-ELTP). NOIRLab is a federally funded center for ground-based, nighttime optical and infrared astronomy. Each semester, NOIRLab facilitates a Time Allocation Committee that is tasked with ranking proposals for telescope time on all NOIRLab coordinated facilities.

The federal government is currently investing in the development of two thirty-meter telescopes: the Giant Magellan Telescope in the Southern Hemisphere and the Thirty Meter Telescope in the Northern Hemisphere. US-ELTP is a joint initiative between NOIRLab and the two observatories to manage the publicly available telescope time for these ground-based observatories. Telescope time is a key resource for doing astronomy and the dependence on (and competition for) observing time as a resource unique to astronomy that does not play out in other disciplines. As a result, institutes, labs and agencies with control over telescope time are uniquely poised to incentivize DEI goals in the same way that federal funders do with financial resources.

The US-ELTP will allocate at least 25% of the observing time on each telescope through an open peer-review process, to be facilitated by NSF's NOIRLab. Proposals for telescope time will be required to submit research inclusion plans as part of their proposals along with research and data management plans. In order to aid astronomers, who are anticipated to be unlikely to be adept at writing or reviewing research inclusion plans, US-ELTP's *Research Inclusion Initiative* developed a Toolkit of Collaborative Practice as a resource for proposers and reviewers. The Toolkit provides astronomers and astrophysicists with best practices on topics like collaboration, cross-institutional partnerships, evaluation and assessment, fostering inclusive environments, mentorship, and recruitment. US-ELTP's RII has emphasized the importance of accountability in furthering DEI goals through research inclusion plans, and as a result, the toolkit also provides a series of metrics for inclusion plan themes so that proposers can assess whether their collaborations are meeting DEI goals.

As of the time of this writing, we are continuing to improve upon the current version of the US-ELTP's Toolkit of Collaborative Practice. An advisory committee of social scientists are reviewing Version 2.0, and based on their feedback we will continue to improve this resource for the community. The Toolkit will be continually updated as a resource to help astronomers align their collaborations with the DEI goals of the community.

9.4 Discussion and Conclusion

In this contribution, I call for GMOs to actively engage in equity change work through mandates that drive DEI in astronomy and astrophysics research. GMOs are uniquely poised to drive change within scientific disciplines because of their control over resources and advanced policy agendas (McCambly & Colyvas 2022). I demonstrate that GMOs have the capacity to drive equity change work by highlighting two successful cases from the field of astronomy: NASA ATP's implementation of inclusion plans as a stipulation for research funding, and US-ELTP's mandate of inclusion plans in order to apply for telescope time. These endeavors highlight how organizations with control over research resources can steer Astronomical collaborations to more closely align DEI goals as a strategy to drive change in the field.

References

Acker, J. 1990, Gend. Soc., 4, 139

Barber, P. H., Hayes, T. B., Johnson, T. L., & Márquez-Magaña, L.10,234 Signatories 2020, Sci. 369, 1440.

Bikard, M., Murray, F., & Gans, J. S. 2015, Manag. Sci., 61, 1473

Bozeman, B., Fay, D., & Slade, C. P. 2012, J. Technol. Transf., 38, 1

Britton, D. M., & Logan, L. 2008, Sociol. Compass, 2, 107

Eaton, A. A., Saunders, J. F., Jacobson, R. K., & West, K. 2020, Sex Roles, 82, 127

Frickel, S., Albert, M., & Prainsaclk, B. 2016, Investigating Interdisciplinary Collaboration: Theory and Practice Across Disciplines (New Brunswick, NJ: Rutgers Univ. Press)

Hackett, E. J. 1990, JHE, 61, 241

Hunt, L., Nielsen, M. W., & Schiebinger, L. 2022, Sci, 377, 1492

Jeon, J. 2019, SSS, 49, 839

Jeske, M., et al. 2022, PLoSO, 17, e0263750

Jimenez, M. F., et al. 2019, NatEE, 3, 1030

Kameny, R. R., et al. 2014, J. Career Dev., 41, 43

Kleinman, D. L. 2003, Impure Cultures: University Biology and the World of Commerce (Madison, WI: Univ. of Wisconsin Press)

Leahey, E. 2016, Annu. Rev. Sociol., 42, 81

Lee, Y. N., & Walsh, J. P. 2022, ST&HV, 47, 1057

McCambly, H., & Colyvas, J. A. 2022, J. Public Adm. Res. Theory, 20, 1

Merton, R. K. 1968, Sci., 159, 56

Mickey, E. 2019, Gend. Soc., 33, 509

Mulvey, P., & Pold, J. 2023, New Astronomy PhDs: What Comes Next https://ww2.aip.org/statistics/new-astronomy-phds-what-comes-next

National Academies of Science, Engineering, and Medicine 2021, Decadal Survey on Astronomy and Astrophysics 2020 (Astro2020).

Porter, A. M., & Ivie, R. 2019, Women in Physics and Astronomy, 2019, Report, AIP Statistical Research Center

Prescod-Weinstein, C. 2021, The Disordered Cosmos: A Journey into Dark Matter, Spacetime, and Dreams Deferred. (London: Hachette UK)

Ray, V. 2019, Am. Soc. Rev., 84,

Reinecke, D. 2021, SSS, 51, 750

Sacco, T., & Norman, D. 2022, BAAS, 54, 028

Smith-Doerr, L., Croissant, J., Vardi, I., & Sacco, T. 2016, In Investigating Interdisciplinary Collaboration (New Brunswick, NJ: Rutgers Univ. Press) 65

Smith-Doerr, L., Fitzpatrick, D., Alegria, S., Tomaskovic-Devey, D., & Husbands Fealing, K. 2019, Am. J. Sociol., 125, 534

Stevens, K. R., et al. 2021, Cell, 184, 561

Walsh, J. P., & Lee, Y. N. 2015, RP, 44, 1584

Wooten, M., & Couloute, L. 2017, Sociol. Compass, 11, e12446

Wuchty, S., Jones, B. F., & Uzzi, B. 2007, Sci., 316, 1036

AAS | IOP Astronomy

An Astronomical Inclusion Revolution
Advancing Diversity, Equity, and Inclusion in Professional Astronomy and Astrophysics
Dara Norman, Tim Sacco and Dorian Russell

Chapter 10

The Scientific Merit of Building and Maintaining a Culture of Inclusion in Astronomy and Astrophysics

Dara Norman

The culture of astronomy and physics must evolve if we want to attract innovative people to the field. Physicists who identify with underrepresented (e.g., BIPOC) groups have long sought to make space for themselves in physics and astronomy. Affinity groups like the National Society of Black Physicists (founded in 1977), the National Society of Hispanic Physicists (1995), and the Society of Indigenous Physicists (2020) have been established to support those who have felt unwelcomed, under-appreciated and dismissed in the fields of physics and astronomy. These groups are established not only for currently practicing professionals, but also to support students navigating their way into a career that can be increasingly hostile the more one succeeds. While many would like to believe these societies to be unnecessary, the myriad of stories from students and early career faculty, as well as the still extremely low numbers of physicists from these backgrounds (see Chapter 1), suggests otherwise. If scientists truly believe that the ability, interest and drive to succeed in fields like physics and astronomy are distributed across the population without regard to gender, race or socio-economic background, then when we don't see those distributions reflected in the professional workforce at numbers similar to their representation in the general public, we know there must be barriers and biases that prevent those interested in joining the field from doing so. Helping to make the opportunities available to those with the interest and drive is the mission of these groups with members linked by a common mission, i.e., affinity organizations.

10.1 Science Culture

Why aren't physicists from minority groups represented in their field at the same proportion as their representation in the broader population? Much of it has to do with the culture of physics and astronomy. Culture refers to the collective customs, institutions, and achievements of a particular group of people. Professional science has its own culture. For instance, science has a distinct Language, represented by equations and how phenomena are described; Traditions that define how and when things are done; Norms that include the expected behavior of others in scientific situations; and Values that acknowledge what is regarded as significant to advancement in the field and by contrast, those activities deemed less important. The field even has its own Art, which may include how talks are organized or how data is taken, as well as methods of analysis. The culture of science encompasses the subculture of Physics, which in turn includes a subculture of Astronomy. Within each of these subcultures, groups will be more or less familiar with various details of practices, norms and values, but the basic tenets are recognizable. As we segment our community into collaborative groups, we bring these traditions, norms, and values along.

One key precept about science culture is that it is not innate, rather it is learned—passed down from mentor to student. Thus, in each successive generation, the language, norms, values, traditions, and arts can evolve. While some of these have evolved over the last 50 years, there are still major barriers and biases that keep the culture of physics and astronomy from being inclusive of specific groups of individuals who wish to participate. Despite all of the advances that have been made as a result of discoveries in physics and astronomy, the full potential of the field will not be recognized without the contributions of those currently being left out. We know this because it has been shown that diverse groups of people, with different heuristics and experiences, are better able to solve complex problems than experts with more common backgrounds and approaches (Hong & Page 2004, Freeman & Huang 2014). The more complex the problem, the more important a diverse team becomes because each member brings their broad experience, background and knowledge to finding answers and solutions to questions and problems. (Phillips 2014).

10.2 Astronomy Cultural Values and Norms

What parts of the astronomy culture need to evolve in order for the community to be more inclusive? An inventory of the field's values, that is, what is deemed to be important enough to incentivize, provides some key insights. In all of science, publications have been relied upon as evidence of the significance of one's scientific work and expertise. However, beyond just one's scientific knowledge, there are barriers to being able to publish. Funding can be a significant barrier, made more difficult by the necessity of changing cost structures for operating open journals that is shifting more cost to the authors (*De Wit* et al. *2018*). Access to data is also among the barriers that are then reflected in the number of publications any individual can point to. Data access can be a pernicious problem because, while many astronomical datasets are open access through archives, there is significant inequity in access to the infrastructure (hardware, software, and human) that allows any individual researcher to take advantage of these open data models. Similarly 'open sky' observing policies allow anyone to propose for public time on telescopes, but if open sky policies are left unchecked, they can mean a perpetuation of the old adage that the 'rich get richer while the poor are left behind', i.e., those who get telescope time are able to make the best cases for more time. In a limited resource situation, those who never get any access are less able to make a compelling case.

Technology and policies are ways that we can start to make progress towards democratizing scientific discovery. In astronomy the number of nights available per year on any telescope are finite... there are no more than 366 nights in any year and with factors like poor weather, maintenance, etc, there are many fewer! However, through data archives, the ways in which investigators can craft experiments is limited only to the ways in which they can imagine using the archival data. Mining data in archives has become easier in the last several decades due to advances in software and networking that allow those users who have limited local computing power the opportunity to reduce, analyze and share their data and workflows from computers located at data centers and observatories. This removes the requirements for specialized infrastructure at the investigator's own location. The use of these 'science platforms' are poised to be game changers for supporting researchers with limited local infrastructure resources by providing better access to archival data and analysis resources.

The best way to understand what researchers at institutions with few resources need is by *asking* them. It seems silly to say, but often advisory committees within the field primarily consist of researchers from large, well resourced universities. This is perhaps not surprising since researchers at smaller teaching /or less resourced institutions may not have the time or internal support to participate on advisory committees that may not be directly beneficial to their home institution. Since there is evidence that smaller/less resourced institutions are also more racially diverse than larger research focused (R1) institutions (Kameny et al. 2014). If those who select committee members are not conscious of the issue, this can also contribute to advisory committees that are less diverse. Furthermore, it is not helpful that there is often no compensation, monetary *or otherwise*, given to committee members for their time and work. The persistent reliance of the community on the assumption that all researchers have the luxury to simply volunteer their time, ensures that those who can't volunteer are often not heard or recognized in places where policies and practices are established. Attempting to do research at a university or college with few infrastructure resources, and possibly distinct priorities with respect to the balance of teaching versus research, is unlike the scientific environment at a research heavy and focused (R1) university. To make space for the broad range of scientists to participate, we need to have a clear understanding of the barriers and biases that are in the way. We can't fully understand what those barriers are if we don't ask those who are experiencing them.

In some cases the barriers may be technical, i.e., a lack of infrastructure resources, as mentioned above. However, in other cases they may be administrative or even social barriers. For instance, when planning scientific meetings and conferences, how often is the timing and the location of the meeting set without even considering the extra burden for potential attendees who are coming from smaller institutions? Unless members of the organizing committee are at such colleges and universities, almost never! Therefore, as we seek to be more inclusive, it is extremely important to take broad advisory access into account.

10.3 The Value of Science Adjacent Work

Engaging the public and educating the next generation of scientists are activities that the community claims are of importance for science. However, often those activities are devalued when it comes to obtaining promotions and other opportunities within collaborations. Furthermore, studies find that underrepresented scientists in collaborations are more likely to be overrepresented in the kinds of work that tend to be devalued (Thiry et al. 2007). Demonstrating that the science community values these activities requires that those who engage in them are made to feel they are integral parts of any science collaboration—the science as well as the outreach (Bauer et al. 2019). Scientists know this work of engaging the public and participating in education is important because the funding to do science depends on the continued support of the general public. If we lose that support and the public is no longer interested in our activities, we cannot achieve our scientific goals. Those who choose to spend their time engaged in these educational and public focused activities are vital to the survival of the field and their contributions should be treated as such. Sadly, this is often not the case (Villalpando et al. 2002).

In a similar way, many of those researchers who choose to focus their expertise on software development for science risk having their achievements devalued by the wider community. While we can all recognize that these efforts are required to achieve science goals, the inherent value of the work of putting together these tools is not always recognized as a significant part of the science result. One reason is because while publications are an important piece of currency for scientific advancement, there are few publications where the work of building the tools, independent of the science they enable, can be independently published and recognized (Smith et al. 2019). Scientists who engage in this work are only becoming more important to the field and the community must find ways to show that this work is valued (Norman et al. 2019).

Large collaborations are increasingly becoming the norm in astronomy research, especially as astronomical research requires increasingly large or timing critical datasets. These lofty science ambitions require telescopes and satellites that are increasingly complex and expensive. As scientists, we often don't appreciate it, but as our collaborations grow, communication, mentoring, and governance also become more complex and difficult to do correctly. Large collaborations are complicated systems that require planning, forethought, and some degree of flexibility to work well. They are important for defining how we do our science—whether it represents an environment of inclusion and equity or one of exclusion and privilege for only some. Building a properly working collaboration requires members to have clear and potentially difficult conversations about governance, policy, and conflicts. It is more likely that if there are clear policies before conflicts arise, and they will arise, can there be the means to resolve them amicably. Those who do the hard work of ensuring that a collaboration is working properly should be valued and recognized for that work because it is essential to the ultimate goal of doing the best science (see Chapter 8 by Bennett).

10.4 How We Do Science Included as Scientific Merit

As I have advanced in my astronomy career, I noticed that despite years of programs and plans, there are still few people from minoritized backgrounds or who are from a similar background as my own in the field. This has been especially true when I look at who is in the leadership of major projects and collaborations. Looking back to the early years of one project, that I will call the Big Astronomy Project (BAP), which I am particularly familiar with, I noticed that from the beginning, this project had a number of excellent features that made it a model in the field. The BAP had plans to build on cutting-edge technology in hardware and software in order to produce great science on a wide range of topics in the field. Furthermore, the BAP was poised to open up brand new areas of scientific discovery, with new and innovative methods for collecting and analyzing data. Even in its very early stages, before any data had been taken, the project had established itself with the broad community of scientists who began collaborating in anticipation of the BAP. The project had already made public outreach and education part of its main goals and thus part of its core funding structure, even in the development phase. Early on, the BAP also included a broadening research participation component that was based on a successful program practiced by another funding agency. The BAP's program included a number of good practices for building cross-institutional collaborative partnerships. Professors from Historically Black Colleges and Universities (HBCUs) and small colleges were paired with BAP staff to work on areas of interest to the project. These professors were able and encouraged to make their student research group (primarily undergraduate students) part of the collaboration. The idea was to build 'pipeline' opportunities for students and successful partnerships between professors and staff such that, when the BAP was finally completed, these partnerships would continue as diverse scientific collaborations. This was a true opportunity to bring research equity and inclusion into the BAP.

However, in my review of the project eight years later, I could identify many of the original excellent aspects of the project—the cutting edge science and technical innovation, the community science groups that had evolved into science working efforts, and the EPO program that continued as part of the broad design and construction effort. However, one area that was no longer part of the mix was the broadening research participation component! So what happened? I conclude that while this broadening research participation program was highlighted by leadership, its support and funding were never made part of the core effort of the BAP—the success of this program was never tied to the reviewed and monitored success of the BAP. This effort to broaden participation within the research areas that the BAP would do was never viewed as part of the scientific merit of the project. Without status as part of contributing to the ultimate success of the project, the program was pushed to the side and then it just disappeared.

If the community of scientists, in all fields, truly want to improve equity, diversity and inclusion in the field, we must make broad participation in research part of how we work as scientists. *How we do our science and how we achieve our scientific goals must be part of how we assess the scientific merit of our projects and our collaborations.* If we don't do our science in an equitable and inclusive way, we will not be doing the best science and thus the scientific merit of our work is diminished. Incentivising this cultural change in how we assess scientific merit will be the next major change we must achieve as we push the boundaries of scientific discovery.

10.5 Consideration of Ethical Behavior in Science Is Not New

In a limited resource environment, there needs to be some way of determining how to allocate those resources. As a result, scientific research is a competitive, as well as a collaborative, endeavor. Whether for funding, observing time on telescopes, supercomputer access, or other resources, the metric by which we judge the benefit of a proposal is through an assessment of its scientific merit. Various review panels will have their own definition of scientific merit, but basically it comes down to assessing what the benefit a proposed project brings to the scientific mission or goals; is the proposed science worth doing and can it be done successfully as proposed.

Whether some kind of science is worth doing is always subjective and will depend on the needs of the field at the time, as well as the interests of the reviewers. Persuasion is the task at hand and the ability to persuade a review panel is not always only about the proposed science (Tomkins et al. 2017). The assessment of whether the proposed work can be done successfully can be judged by multiple metrics. There are the technical aspects of the proposal—is it feasible? Is the design of the experiment sound—will the outlined goals of the experiment or project be fulfilled if the research is done in the proposed manner? Also, will the proposed science be accomplished in an ethical way?

The advancement of science has not always been done in an ethical way. Most physical scientists will be most familiar only with ethical issues around plagiarism or the falsification of data. Social and medical scientists are familiar with more strict expectations around the design and implementation of experiments because of requirements for getting Institutional Review Board (IRB) approval when working with human subjects. However, the need for ethical behavior when conducting science need not be confined to the subjects of experimentation, adhering to ethical behavior should also include the treatment of those doing the science (Bennett in Chapter 8, Resnik 2020).

The ethical treatment of the scientific workforce has become more of a concern in recent years. Examples, including the implementation of codes of conduct among scientists and within collaborations, have begun to be normalized. Anti-harassment and anti-discrimination statements are more common expectations at meetings, at regular gatherings of scientists, and in scientific workplaces. These actions demonstrate that scientists recognize the importance of ethical considerations for doing their best work and that workplaces that allow bad behavior to go unchallenged are not pleasant or productive places to engage in scientific discovery or innovation. Therefore, it is reasonable that as we review the scientific merit of a proposal or a project, we consider all aspects of whether and how the science will be accomplished, including that the work be done in an ethical way.

The recent requirements, in some astronomy and physics sectors, for considering 'Inclusion Plans' in the review and assessment of proposed research is an opportunity to consider 'how science is being accomplished' as part of a proposal's scientific merit. As discussed above, this is not a new idea, however rarely before in astronomy and physics have we made this additional, non-technical question of 'how science will be done' explicitly part of the review of an initial proposal. Introducing such plans as a requirement for proposals provides an opportunity to have primary investigators and their team think, in a serious way (because it will be judged), about the structures and practices of their teams, groups, and collaborations. Furthermore, these plans enable the possibility of a way to hold PIs accountable if they fail to 'deliver' on the work that they proposed to do. Currently, proposal calls for funding from NASA and the Department of Energy[1] have begun to explicitly require inclusion plans for science proposals. Plans for the US Extremely Large Telescope Program[2] will extend such requirements to NSF resources like ground-based telescope observing time. It is important for the community to recognize and acknowledge that these requirements are not only reasonable for science proposals, but also that requiring a commitment to ethical behavior as part of the scientific merit of the work that scientists do is long overdue in the physical sciences.

10.6 Building and Maintaining a Culture of Inclusion

Diversity, Equity, Inclusion and Access need to be part of *all* the work that we do in astronomy and astrophysics, from how our research and analysis is done, to how we do our teaching and mentoring. These must be a part of how we design and execute experiments, as well as how we share our discoveries with the rest of the world. It is only through these actions that the Culture of Inclusion can be experienced by all parts of the community. We include the principles not only because it improves our science, but also because it is the right thing to do. Maintaining a culture of science that supports DEIA will be easy once we incorporate it into all how we do and share science. Considering the ways in which DEIA impacts all areas of our science and making those part of how we assess the scientific merit of our work is key to building and maintaining a culture of inclusion in astronomy and astrophysics.

---

[1] In 2020, NASA introduced the idea of 'inclusion plans' for science proposals, i.e., https://science.nasa.gov/researchers/inclusion/. Requirements for such plans has spread to programs within DOE, which require PIER plans, https://science.osti.gov/sbir/Applicant-Resources/PIER-Plan
[2] See Sacco article in Chapter 9.

References

Bauer, A., et al. 2019, BAAS, 51, 130

Freeman, R. B., & Huang, W. 2014, Natur, 513, 305

Hong, L., & Page, S. E. 2004, PNAS, 101, 16389

Kameny, R. R., et al. 2014, J. Career Dev., 41, 43

Norman, D., et al. 2019, BAAS, 51, 24

Phillips, K. W. 2014, SciAm, 311, 42

Resnik, D. B. 2020, What Is Ethics in Research & Why Is It Important? https://www.niehs.nih. gov/research/resources/bioethics/whatis/index.cfm

Thiry, H., Laursen, S. L., & Liston, C. 2007, JWM, 13, 391

Smith, A., et al. 2019, BAAS, 51, 52

Tomkins, A., Zhang, M., & Heavlin, W. D. 2017, PNAS, 114, 12708

Villalpando, O., & Delgado-Bernal, D. 2002, The Racial Crisis in American Higher Education: Continuing Challenges for the Twenty-First Century, ed. W. A. Smith, P. G. Altbach, & K. Lomotey (Albany, NY: SUNY Press) 243

Part IV

Astronomy Activism Driving Equity and Inclusion

Introduction to Part IV: Astronomy Activism Driving Equity and Inclusion

The goal of Part IV is to showcase how astronomers are thinking about and engaging in activism in their field. In recent years, astronomers and astrophysicists have increasingly pushed for greater diversity, equity, and inclusion (DEI) within the field. However, our professional work does not exist in a vacuum; rather, we do our work within a broader "institutional environment" of overlapping cultural norms and values, and formal legal and bureaucratic structures, that shape people's on-the-ground experiences. Among scientific disciplines and within society more broadly, the dynamics of these institutional environments tend to uphold the status quo and, by default, reinforce inequalities on the ground. If institutions tend to reinforce the status quo, how can effective change be fostered within astronomy?

The first contribution to this section, *"The Way that the COVID Pandemic Amplified Inequalities in Astrophysics"*, highlights how several long-term DEI efforts in astronomy were disrupted by the COVID-19 pandemic. The piece argues that, while the pandemic has some marginal gains for inclusion, it mostly undermined the "collaborative project" of DEI work and exacerbated long-standing inequalities in astronomy through unequal impacts on already marginalized groups. Then, *"Queering Physics through Creative Conflict"* draws on the author's experiences in social justice-oriented programs in physics to demonstrate how underrepresented scientists can work together to actively facilitate a more inclusive environment for themselves, and also—on a deeper, cultural level—to disrupt stereotypes of who can be a physicist or what a physicist should look like.

Finally, *"Keep Calm and Carry On...Nevermind the Backlash!"* highlights how, alongside the many gains toward DEI goals in astronomy and other STEM disciplines, there is a growing, organized resistance to DEI. The contribution highlights the benefits of DEI to scientific communities as well as scientific research. The piece argues that astronomers should continue working towards DEI goals in the face of resistance to these policies, and to stand up against efforts looking to undermine ongoing efforts to make STEM more inclusive and equitable. Taken as a whole, Part IV encourages astronomers to actively strategize together in the face of inequalities and to forge new paths toward meaningful change in STEM.

An Astronomical Inclusion Revolution
Advancing Diversity, Equity, and Inclusion in Professional Astronomy and Astrophysics
Dara Norman, Tim Sacco and Dorian Russell

Chapter 11

The Ways that the COVID Pandemic Amplified Inequalities in Astrophysics

Jarita Holbrook

11.1 Introduction

Stay at home regulations implemented due to the COVID-19 pandemic impacted the lives of everyone. For astronomers and astrophysicists, staying at home meant switching their lives, which were normally rich with travel to conferences, to observatories for data collection, and simply to work for departmental meetings, team meetings and teaching, to having all of these occur online. The switch to online work was not seamless. Some universities were better prepared than others. In South Africa and other countries in the global south, many students could only access the internet on their cell phones and data rates were prohibitively expensive. In response, South African universities had to organise funds and negotiate rates with cell phone providers to ensure that their students were able to fully participate online.

The pandemic also impacted automated telescopes such as the Sloan Digital Sky Survey (SDSS) (Abdurro'uf et al. 2022). The Apache Point Site atypically operated during the Northern Hemisphere summer months to make up for being shut down for nearly 2 months in 2020. The SDSS Las Campanas site was shut down from March to October 2020 (Santana et al. 2021), and like Apache Point, extra nights of observing were added once reopened. However, shifting observations by months meant observing a different part of the sky than originally planned, thus investigators had to replan. Other observatories, such as those on Mauna Kea, that previously offered in-person and remote observing options, shifted to entirely remote observing, though the telescope operator was still on-site.

In addition to the issues of moving online and data collection, pandemic lockdown of schools and daycares also had an unequal impact on families with school-aged children. Astronomers with such families struggled to complete astronomy-related tasks during business hours given the added childcare duties and even having to teach their children during these hours. Thus, it was often necessary for these astronomers to shift their work hours to later. Many experienced mental and physical exhaustion after tending to their children much of the day. Under those conditions, these astronomers found it difficult to work during the evenings. Astrophysicists anticipated such problems and warned that this differential impact of COVID on work would show up in CVs adversely (Venkatesan 2020), thus there could be long-term impact on astronomy and astrophysics as those with children would appear to be less competitive and thus would be less able to secure employment in the future. Additionally, given that those without children found the lockdown periods peaceful with less distractions such as no longer having to commute, many were able to get more research done and submit more articles for publications. The contrast between these two broad experiences reinforces Venkatesan's concerns.

Within large collaborative groups, in theory, the group members would include those with and without children, and with and without extra caring duties. Thus, the impact of COVID could be masked because those with more time on their hands could potentially carry the load of analysis and writing for the group.

ASTROMOVES is a qualitative research project using interview-based inquiry to explore the lives and careers of astrophysicists after their PhD (Holbrook 2019). The goals of the project focus on how people make decisions about their careers for each of their career moves post PhD, how identity (i.e., intersectional identity[1]) plays a role in their decisions, and if the academic/work environment is important.

Conclusions about these are to be used to fashion suggestions for retention policies, which is of importance for collaborative teams. The project was started during the pandemic and the scientists spontaneously included COVID in the interviews, thus the original interview questions were broadened to include questions about COVID. These ASTROMOVES interviews offer a different way of exploring the impact of COVID-19 and the inequalities already present in astrophysics.

11.2 Data Collection

The first ASTROMOVES interviews took place in late 2020 and to date 37 interviews have been collected. As with the lives of the astrophysicists and scientists interviewed, the COVID pandemic changed the data collection protocol for the project. Rather than in-person interviews, the interviews were done online. Each person was interviewed for between one and two hours about their changes in position since receiving their doctorate degree and their decision-making. To be eligible for the study, astrophysicists and scientists in adjacent subjects had to meet two criteria: (1) they had to be European, gotten their doctorate in Europe, or held a position in Europe at some point during their career, and (2) they had to be at least two positions past their doctorate, which is estimated to be at least five years past their doctorate.

Of importance was to explore how intersectional identities influenced careers and career decision-making, thus the goal was to include women, LGBTQIA+ people,

---

[1] Intersectional Identity refers to how different axes of identity, like race or gender, are "mutually reinforcing sites of power relations", which shape the personal and professional experiences of people from different social backgrounds (Scott 2017: 370; Collins 1990)

people with disabilities, people of non-European heritage, and Eastern Europeans. All scientists that participate in the project are volunteers. To recruit scientists a combination of targeted sampling and snowball sampling was used. The LGBTQIA+ astrophysicists that were interviewed were identified through the astrophysics "outlist" (Mao & Blaes 1998) and others joined through recommendations from people that knew about the ASTROMOVES project. Given that this project is a Marie Skłodowska Curie Actions Fellowship, current and previous MSCA fellows in astrophysics and adjacent sciences were emailed and invited to participate. Snowball sampling was used, in which scientists interviewed were asked to recommend other scientists to interview, or after seeing an ASTROMOVES presentation or article, scientists volunteered to be interviewed.

11.3 Analysis

Each participant was assigned a Hawaiian name to preserve their anonymity. Hawai'i is the authors' birthplace, thus their familiarity with Hawaiian names; but subversively, using Hawaiian pseudonyms is meant to remind astrophysicists of their debt to the Hawaiian people for allowing us to use their sacred mountains for astronomical observations. Some participants agreed to not be anonymous and will appear in a future film on the project, but for the purposes of articles the decision was made for everyone to be quoted using their Hawaiian name instead of their real name. Every effort has been made to present quotes without identifying characteristics through which identity can be inferred. All interviews were transcribed, and then analysed using the qualitative analysis software NVIVO(*NVIVO* (version 12.7.0) 2021). The discussions about the pandemic were coded for content and then for emotions.

11.4 Demographics

The ASTROMOVES interviews are unique because of the intentional inclusion of diverse scientists. Table 11.1 shows the gender and sexuality diversity of the scientists. In describing themselves, they often gave gender and sexuality information together, thus they are presented together here. Though neutral pronouns are used throughout in an attempt to mask identity, some of the findings are specific to women or to LGBTQIA+ people.

The ASTROMOVES interviewees are ethnically diverse (see Table 11.2) but predictably are majority of European descent. In the analysis, there was no discernible pattern connecting the COVID experiences, emotions, and comments to ethnicity.

Table 11.1. Gender and Sexuality of the ASTROMOVES Participants

Heterosexual Males	15	41%
Heterosexual Female	13	35%
LGBTQIA+	9	24%
Total	37	100%

Table 11.2. Ethnicities of the ASTROMOVES Participants

African Descent	3
Asian Descent	3
European Descent	27
Middle East Descent	2
Mixed	2

Table 11.3. The Career Age of the ASTROMOVES Participants

Career Age	Heterosexual Females	Heterosexual Males	LGBTQIA+ members
>20	3	5	1
10–19 years	5	7	4
<10	5	3	4
Total	13	15	9

Thus, the scientists' ethnicities show the diversity of the ASTROMOVES participants but were not relevant for the analysis.

The Career Age of the scientists is shown in Table 11.3. Rather than age, the Career Age, which is calculated from the year they were granted their PhD, is a better measure to use when discussing the career decision-making that is central to ASTROMOVES. The average career age is 17 years (± 12 years) for Heterosexual Males, 15 years (± 9 years) for Heterosexual Females, and 13 years (± 7 years) for LGBTQIA+ members; thus, the ASTROMOVES LGBTQIA+ members are slightly younger on average. For this analysis, career age was important because it is more likely that they have stable employment at higher career ages, thus minimizing the pandemic's impact on career navigation.

11.5 Working Online or in Person

I mean a lot like, of course [the pandemic has] affected everybody's life. Profoundly and so I'm certainly no exception to that. Yeah, I mean, before, before COVID, I would [...] travel a lot... I have friends in many different places. I like to think of myself as being very research active. So, I get lots of invites to conferences and seminars. And there's lots of interesting things I want to attend in that way. My family lives abroad, etc. So, a big part of my life was... travelling around to do all these things. And, you know, that is all completely stopped now. So that's disappointing. I've managed better than I might have guessed. If some-one had told me a year ago that something like this was going to happen. In that regard. Thanks to...video calls and things like that. But yeah, it's not, it's not great. In terms of other aspects, so...I'm by myself.

So, I have...no human interaction whatsoever and haven't for nine months. And that's not great. You know, aside from again through the computer. So again, I'm managing okay for now because there's an expectation that this will end eventually. Um, but yeah, I mean, you know, it's a pretty profoundly unpleasant experience.

—Maka'ala (2020)

Being able to move online to connect with students and complete work, many would consider a positive. However, Maka'ala and Honokakua were among those that preferred interacting in person, thus did not enjoy the online experience.

Honokakua: We work with computers, we can have meetings on Zoom, even though I don't really like this type of interaction. I prefer to meet people in person and to discuss with colleagues and so on, in person, but it was okay.

Haoa is the director of an astronomy facility that was commissioning a new instrument, the timeline of which could not be adjusted. Thus, Haoa was required to work in person throughout the pandemic. There were safety precautions in place, such as 50% occupancy of buildings and extra cleaning protocols. Nonetheless, there was concern about safety and being put at risk because of the project.

Given the cutting-edge technology that was a part of being a national facility, Kilauea was seamlessly able to shift from working in person to working online. The move online was beneficial because Kilauea no longer had a long daily commute, which they found exhausting. In comparison, Kilauea had more intellectual energy and time that led to publishing more since the pandemic began. Also, Kilauea prefers working remotely and had requested working remotely but had been denied, that is until the pandemic made it possible.

The response to being online was mixed with people like Kilauea benefitting from it, but many others not enjoying it. None of the participants spoke of situations like that in South Africa, where they were unable to afford to go online or they did not have access to the internet.

11.6 Job Security/Insecurity

The end of my most recent research fellowship then three months of unemployment and it was pretty harrowing: I'VE NEVER BEEN UNEMPLOYED BEFORE!... I've actually got to [the] interview stage for a couple of jobs and, and, they were just unceremoniously called off... months later they called me back and said, 'We'd like to resume the interview process with you.'

—Kamea (2020)

What about job searching during COVID-19? An already tight job market got tighter as job searches were frozen, job offers retracted and the closing of advertised positions. This research suggests that underrepresented astronomers may face an especially precarious job market that their white male counterparts may not face.

Of the 37 people interviewed, 8 had their jobs impacted by the COVID-19 pandemic. Kamea, Laka, Lele, Wena, and Hilo were on the job market when the pandemic began. Though Kamea was shortlisted for a few positions, most of those searches were suspended. One institution resumed their search after a four-month hiatus. Thus, in response to the pandemic, freezing the process of filling advertised positions, including the cancelling of positions, led to Kamea being unemployed for many months. Kamea had to apply for and receive public assistance to cover rent while waiting for another position.

Wena had a job offer that was frozen for a month. During that time of waiting for confirmation and then waiting for their first paycheck, Wena had to live on their savings.

For Lele, the process was much longer because of the pandemic. They were not the first choice for two positions, but both jobs eventually were offered to them. Lele accepted a position in the USA, then the political situation changed for obtaining work visas in the USA. The second position came through and Lele decided to decline the first position and accepted the second position. Additionally, the second position allowed Lele to stay in the same city rather than move to another country.

Laka commented that there were a lot fewer positions advertised due to the pandemic, which made the job search process more difficult.

Hilo was given a job offer, which they declined, but reflected that the way the offer was handled must have been due to the pandemic. In particular, the offer was made with less than a week given for a response and a non-negotiable start date 30 days later, both unusual. Hilo would have had to quit their current job and move to another country in that short timeframe. Hilo later found out that it was the institute that handled the offer badly, not the scientific team. Relatedly, Holualoa recommended that postdocs on the job market during the pandemic negotiate for what they need and acknowledge to themselves that their needs will be different because of the pandemic.

Hekela: Yes, yeah, so, I've been able to, to, to get, to get _____ work done. Still, not as much as I had before pre COVID but...

Hekela, who owns their own business, was impacted by COVID-19 specifically by the different levels of lockdown in their country. Hekela's situation is complicated by having to share childcare duties as described in the Extra Duties. Work was not stopped by the pandemic, rather it was slowed down by not attracting new customers and having to find time for the existing customers.

Hema's job was up for renewal, which caused them some stress. However, Hema's contract was renewed for another three years. This was for a job connected to working for a national research facility outside of academia.

Hanohano has been working on a series of contracts at the same university post-PhD, however over the next year, they will need to reduce their hours because there isn't enough funding to pay them full time. Previously, their university and their department were good about bringing together funding for a full-time salary. Hanohano thinks the pandemic has put strain on the university and department finances as well as on the people handling finances, leading to their future part-time status.

Summarising, of the eight people that had their jobs impacted one person had to get on public assistance, another had to live off their own savings until the next position came through, and a third anticipates having a reduced salary within the coming year. Importantly, the participants cited in this section include women, non-heterosexuals and the disabled; none of these are able-bodied heterosexual men. This may point to a precarious job situation connected to diverse and intersectional astrophysicists. Collaborative groups need to have an awareness about these issues so that they can preempt team members becoming unemployed or having to rely on their own savings as they are waiting for their next position.

11.7 Wailoaloa's Interview

Wailoaloa expressed unique views on how the pandemic changed how they are thinking about navigating their career and how the pandemic impacted their romantic relationship.

For example, Wailoaloa says the following:

> For me it's not so much what the... prestige of a place is. Maybe at some point when I was much younger that was an important reason. But for me, now, it is much more actually about the place than...I mean culture, local culture than actually the university. So, if I were to find, you know, some small University in a city that I really appreciate to live in, I wouldn't mind at all being there. Is particular nowadays, after the pandemic, that we are so well interconnected... there is no shortage of talks to attend and there is no shortage of ways of communicating with global researchers elsewhere. So, I think that would be completely fine.
>
> —Wailoaloa (2020)

Thus positively, Wailoaloa sees the newly interconnected world as offering the opportunity to prioritize living in a desirable town or city instead of focusing on working at prestigious institutions. Wailoaloa was doing a postdoc in Europe while their partner was in North America. Their relationship was impacted negatively by the onset of the pandemic:

Wailoaloa: Things were not clear what would happen [in our relationship], but then the pandemic hit, and then with the pandemic everything became very complicated...In terms of planning...simply seeing each other. In terms of like decision making, so life took another turn...

Jarita Holbrook: So, if COVID hadn't happened, you think that relationship might have survived?

Wailoaloa: It's unclear to me. It's hard to say. But, for instance, we know that... Last summer, we were supposed to see each other more than once, and then couldn't see each other for actually the whole year, basically, and part of the main reason was about travel restrictions.

These were two unique ways, romantically and in terms of future career planning, that the pandemic impacted and has influenced Wailoaloa's life. The issue of finding two jobs in the same region for academic couples or having a spouse that cannot move is not new, but additional tension was added due to COVID restrictions. Physically being away from their partner caused the split in Wailoaloa's case, but emotions such as loneliness due to being away from their emotional support system or due to isolation also could be part of it. Collaborative groups tend to prioritize finding spousal positions only when the person being hired will be permanent, but are less active when people are considering temporary positions such as postdocs. COVID has shown that being away from family (emotional support) can negatively impact work performance, thus such policies should be revisited.

11.8 Extra Duties

Several researchers have reported upon the impact of childcare and eldercare duties on the careers of primary caregiver academics (500 Women Scientists et al. 2020; Gewin 2020; Minello 2020; Venkatesan 2020; Fisher & Ryan 2021). The ASTROMOVES participants with children were no exception. Hekela has a child in nursery school. During the strictest level of lockdown childcare facilities were closed.

"It certainly did during lockdown, the spring lockdown when nurseries were shut and our child was home. I basically stopped all ____work, except for maybe like a couple hours a week. I did, sort of pick up my second of two regular clients during the lockdown period, but that meant a lot of getting up at 5 am and putting in two or three hours of work before… and [Spouse] has been able to be pretty accommodating to give me a couple of hours, a few days a week, in the middle of the day." —Hekela (2021)

Hekela's choice of words indicate that their career is subordinate to their spouse's career as well as indicates that Hekela is the primary caregiver for their child. Under less strict lockdown conditions, the nurseries were open. With their child out of the house, Hekela was able to dedicate more hours to their clients; though they have yet to reach pre-COVID levels of work.

Honokakua: So, it was difficult at the beginning of the pandemic, in the sense that daycare was closed. I was taking care of my daughter. So, of course, a little bit of reduction in productivity. Then I was trying to take her during the day, trying to catch up on work at night. So, I had some bad… to say…it was not a lot of sleep and this kind of thing, so it was quite tough.

Honokakua was in a similar position to Hekela because their spouse is a medical professional so was working outside of their home during the pandemic.

Lele had parental leave and extra duties with a toddler while trying to get back to doing full-time research. This was made harder by the pandemic, which meant reintegration into work was done online rather than in person. Thus, Lele felt that they never really returned to work. Another thing Lele noted was that there is a lot of information that is conveyed via informal conversations which just cannot be recreated online.

Kilauea, who benefitted from moving online and published more articles, is married but does not have children, thus no extra care duties. Similarly, Maka'ala, who doesn't have children, did speak of being more productive during the times they were not teaching:

I'm not happy about how this has affected my, you know, my life and my ability to be with people that I care about…in terms of my productivity, it has had no effect or even a positive effect, because all the things like travel and hanging out with friends or, you know, just whatever that I would do, I don't have any more. So, I have fewer distractions.

—Maka'ala (2020)

The scientists quoted in this section are male and female, and some are not-heterosexual; the determining factor is children or not. Other scientists spoke of the many interruptions that occurred during the day because they were at home with their families, thus having uninterrupted time was shifted to early morning or night. None of the ASTROMOVES participants spoke of having eldercare duties. Thus, the ASTROMOVES participants reflect what other researchers are reporting: that having extra caring duties is impacting their productivity whether that means getting new clients or publishing articles. Having flexible deadlines is one way that collaborative groups can accommodate those with extra caring duties. Another is to temporarily assign people to tasks that are less thought-intensive but still need to be completed and support the collaboration.

11.9 Mental Health

Finally, there is the concern about mental health (Editorial Team 2020).

Hema spoke at length about being more extroverted and therefore really had a need to interact with other people daily. Thus, Hema opted to continue to go into their office as well as to meet with people after hours with social distancing. Hema felt this was necessary because of living alone and therefore needed to combat loneliness.

Kilauea was happy about working from home, but suffered what they considered normal sadness and depression over the high number of COVID-related deaths and the suffering of COVID survivors ("long COVID").

Maka'ala, quoted earlier, mentioned that they had not been in contact with other people for nine months and concluded that it was a "profoundly negative experience".

The ASTROMOVES analysis of mental health issues is ongoing. Since COVID began, many universities have recognized the increase in negative mental health and in response have emphasised those mental health facilities available to faculty, staff and students. Similarly, collaborative groups should make their team members aware of which mental health support facilities are available to them.

11.10 Conclusions

Though Kilauea spoke of the benefits of the COVID pandemic in terms of moving online, most of the scientists interviewed for ASTROMOVES saw few benefits. Inequalities in terms of women and men are apparent in terms of extra caring duties; however, the ASTROMOVES scientists had male primary carers that spoke of their careers being slowed because of their extra duties. Still, the ASTROMOVES carers were primarily women. Team leaders should consider using flexible deadlines and temporary assignments to allow those with extra caring duties to remain active team members.

Women, diverse scientists and disabled scientists experienced more job insecurity due to the COVID pandemic in comparison to the heterosexual men of European descent in the ASTROMOVES population. Can this be generalized to the larger astrophysics community? Unclear, but the leaders of collaborative groups need to be aware and take pre-emptive steps to ensure the job security of their diverse team members. The focus on individual scientists, such as Wailoaloa, add to the possible experiences that happened due to the pandemic. Wailoaloa's experience reminds us of the long-standing issue of finding positions for the spouses of the people being hired. In terms of COVID's impact on mental health, there appears to be a pattern not differentiated into males and females, as to those that are single and those that are not, in terms of loneliness and depression due to isolation. Collaborative groups need to ensure that their team members are aware of mental health facilities available to them. ASTROMOVES, though different from other studies due to the strong LGBTQIA+ component, reflects the findings of other researchers in the ways that scientists and academics lives have been impacted by the COVID-19 pandemic, while adding additional impacts previously unrecorded.

References

500 Women ScientistsJedd, T., Goldman, G., et al. 2020, Scientific American Blog Network, 2020 https://blogs.scientificamerican.com/voices/scientist-mothers-face-extra-challenges-in-the-face-of-covid-19/

Abdurro'uf, Accetta, K., Aerts, C., et al. 2022, APJS, 259, 35

Collins, Patricia Hill 1990, Black Feminist Thought (Milton Park: Routledge)

Editorial Team 2020, NatAs, 4, 431

Fisher, A. N., & Ryan, M. K. 2021, Group Processes & Intergroup Relations, 24, 237

Gewin, V. 2020, Natur, 583, 867

Holbrook, J. C. 2019, Proc. IAU, 15, 286

Kamea, 2020, Kamea ASTROMOVES Interview, Interview by Jarita Holbrook. Zoom Recording Audio, Video and Transcript

Lele, 2021, Lele ASTROMOVES Interview, Interview by Jarita Holbrook. Zoom Recording Audio, Video and Transcript

Maka'ala, 2020, Maka'ala ASTROMOVES Interview, Interview by Jarita Holbrook. Zoom Recording Audio, Video and Transcript

Mao, Y.-Y., & Blaes, O. 1998, Astronomy and Astrophysics Outlist, https://astro-outlist.github.io/90s/

Minello, A. 2020, The pandemic and the female academic Nature World View

NVIVO (version 12.7.0) 2021, Mac. English. QSRInternational, https://www.qsrinternational.com/nvivo-qualitative-data-analysis-software/home

Santana, F. A., Beaton, R. L., Covey, K. R., et al. 2021, AJ, 162, 303

Scott, J. 2017, Oxford Dictionary of Sociology (Oxford: Oxford Univ. Press)

Venkatesan, A., Bertschinger, E., Norman, D., Tuttle, S., & Krafton, K. 2020, Women in Astronomy (blog), https://womeninastronomy.blogspot.com/2020/07/the-fallout-from-covid-19-on-astronomys.html

Wailoaloa, 2020, Wailoaloa ASTROMOVES Interview, Interview by Jarita Holbrook. Zoom Recording Audio, Video and Transcript

AAS | IOP Astronomy

An Astronomical Inclusion Revolution
Advancing Diversity, Equity, and Inclusion in Professional Astronomy and Astrophysics
Dara Norman, Tim Sacco and Dorian Russell

Chapter 12

Queering Physics through Creative Conflict

Jessica Esquivel

In, near, and out of the box

Change means movement. Movement means friction. Only in the friction-less vacuum of a nonexistent abstract world can movement or change occur without that abrasive friction of conflict.

—Saul Alinsky Rules for Radicals, Founder of Community Organizing

12.1 Introduction

My name is Dr Jessica Esquivel and I'm an associate scientist at Fermilab. In 2020 I participated, organized, or led upwards of 20 initiatives, in the midst of a pandemic, in the midst of civil unrest in our country, and in the midst of grieving the loss of my dad. All of them focused on spotlighting the lack of just and equitable spaces that Black, Indigenous, people of color, gender-minoritized, and queer individuals have to move through every day. My aim is to change the physics spaces we work in so that we aren't uncomfortable, but that requires uncomfortable conversations, introspective discussions, and the dismantling of a system that continually oppresses me and folks that look like me.

We live in a world where race-based stressors, from individual to systemic, are prevalent, and our academic ivory towers don't shield us from harm. In fact, STEM fields exacerbate these stressors. Research shows that dealing with the constant bombardment of these stressors leads to declining mental and physical health (e.g., Bork & Mondisa 2022) and an overall decline in self-esteem and self-efficacy. Imposter syndrome (e.g., Robinson et al. 2016; Chakraverty 2020) is a symptom of race-based stressors, not a separate internal or stand-alone mentality (e.g., Bernard

doi:10.1088/2514-3433/ad2174ch12
12-1

2020; Stone 2018). Even though logically I know I've moved mountains and shattered glass ceilings to become a particle physicist, I constantly question my worth as a scientist, question if the accomplishments I've garnered were out of pity, or if my intersectional identities are just palatable enough for the white supremacist patriarchal STEM landscape. This constant doubt of abilities is a race-based stressor I deal with constantly and my goal is to turn down the cacophony of negativity for the next generation of Black Indigenous and People of Color (BIPOC) STEM scholars.

Social justice has always been embedded in my understanding of physics. In college, you could not major in physics unless you took a course titled "fiesta physics" each semester. The course focused on community engagement, science communication, and bringing awareness of the opportunity to work in STEM fields to historically marginalized and low-income populations. However, at the time, I wasn't aware that's what I was doing. The intersections of social justice and STEM are still part of how I do physics. Now, I'm intentionally focused on transformational change and using my voice to shed light on deeply rooted injustices in my field. I'm an American Association for the Advancement of Science (AAAS) IF/THEN Ambassador, which focuses on being visible role models in STEM to middle school girls. I'm on the Steering Committee of APS Inclusion, Diversity and Equity Alliance (IDEA), whose mission is to create a social movement to support community organizers at physics institutions across the country, who are working to transform the culture of physics. I'm a member of the Change—Now collective whose calls to action are directed at creating transformational change at Fermilab. Lastly, I'm a co-founder of Black In Physics, a grassroots movement focused on highlighting the contributions of Black physicists and facilitating community building amongst ourselves as we are often the only ones in our respective institutions. All of these organizations have the same vision in mind—that physics, and STEM more broadly, can and should be accessible to all who wish to study it. Being a part of these movements is how I'm queering who is a physicist, i.e., disrupting ideas of what a physicist looks like, and what a physicist does. I will expand on each of these initiatives, and juxtapose the various forms of collective organizing within, adjacent, and outside of traditional institutions.

12.2 AAAS IF/THEN: In the Box

The AAAS IF/THEN organization is a testament to *thinking* outside of the box, and providing the necessary resources to be successful, even while residing within the traditional box of institutional constraints. In partnership with Lyda Hill Philanthropies, a coalition of over 120 women in Science, Technology, Engineering, and Mathematics (STEM) were chosen for the inaugural IF/THEN ambassadorship. This ambassadorship was prompted by the findings of the *Scully Effect Report* by the *Geena Davis Institute on Gender in Media*.[1] It was shown that the influence of the character Dana Scully from *The X-Files* had a positive impact on girls' and women's perceptions of STEM. The report highlighted the lack of representation of women in

---

[1] https://seejane.org/wp-content/uploads/x-files-scully-effect-report-geena-davis-institute.pdf

STEM in media and also underscored the importance of girls' and women's perception of STEM due to their underrepresentation in media. Armed with this data, Lyda Hill Philanthropies[2] threw away the traditional ways of increasing the representation of women in STEM and drew up a new rule book. In partnership with AAAS, Lyda Hill Philanthropies developed a framework that leaned into the importance of representation of women in STEM in the media and elevated the platform of the IF/THEN ambassadors by using a talent agency model. With the lofty goal of activating a culture shift in STEM, the IF/THEN ambassadorship supports current women in STEM by providing grants and media opportunities to shift them to high-profile role models, as well as focusing on intentional media development for the core focus group of young girls.

The most impactful product developed from the IF/THEN ambassadorship is the *IF/THEN collection.* It consists of the largest free resource of its kind dedicated to increasing access to authentic, relatable images of real women in STEM and consists of thousands of photos, and videos. The collection intends to shift cultural perceptions of who works in STEM and inspire middle school girls to pursue STEM careers by exposing them to content that presents STEM as exciting and relatable. The collection provides museums, educators, non-profit organizations, parents, and students with high-quality STEM content for free. The collection also acts as a resource to empower IF/THEN ambassadors by providing resources to help them better share their stories.

The IF/THEN coalition was interviewed, photographed, and trained in media communications and social media skills. We were 3D scanned to create the *IF/THEN exhibit,* which consisted of full-size statues of all 120 ambassadors, that has been experienced by more than 3 million people. The IF/THEN exhibit is a visual celebration of modern women in STEM and it features the most women statues ever assembled together. As the representation of women in STEM is a core principle of the IF/THEN coalition, developing the IF/THEN exhibit sat squarely within that goal. A 2016 study found less than six statues in the top 10 largest U.S. cities were of women.[3]

A side effect of bringing phenomenal women in STEM together is the strong sense of community, collaboration, and support the ambassadorship created. The first in-person conference we had was in October of 2019 and many of us couldn't believe we were in the same room with over 120 amazing women in STEM. Imposter syndrome and doubt filled the air but like a forcefield, the support and community that emerged protected this group of newly minted high-profile role models. We created a counterspace[4] as a by-product of the main goal of the ambassadorship, and it made the group so much stronger.

---

[2] https://www.lydahillphilanthropies.org/

[3] https://empowerment2026.org/national-historic-american-women-statue-initiative/

[4] From Isler et al. (2021): "we draw on the literature about counterspaces. These shared spaces allow one to identify similar experiences within the STEM culture and find a sense of camaraderie, belonging and acceptance. When these spaces are specifically designed for students who have been marginalized in STEM spaces to counter those experiences, they are called counterspaces" https://www.mdpi.com/2313-5778/5/1/8

12.3 Change—Now: Near the Box

Over the course of 2020, we were faced with a pandemic that disproportionately affected BIPOC communities, while also being bombarded with news cycle after news cycle of so many Black individuals murdered by cops. In May 2020, in the wake of George Floyd's death, the Black Lives Matter (BLM) protests gained international attention. BLM was created in 2013 after the killing of Trayvon Martin and the acquittal of his murderer and gained national attention after the murders of Michael Brown and Eric Garner. George Floyd's murder case mirrored that of Eric Garner's case as both defense tactics were to, in essence, blame Black bodies for not being strong enough to take police brutality.

These deaths, the collective grief, and the continued trauma of Black individuals —including Black physicists because we do exist—were taking a toll on our mental health and it wasn't being talked about openly in physics settings. For me, it was easier to explain away this deafening silence by focusing on the broad stigmatization around mental health in general. The harsh reality that physics specifically didn't care about the well-being of its Black practitioners came to a crescendo in 2020 when both the pandemic and the BLM protests collided. Week after week from March-June of 2020, I was receiving emails openly discussing and naming potential mental health issues due to COVID-19 and teleworking. Week after week my superiors were discussing ways to address mental health concerns. Mental health wasn't a taboo subject once it started affecting White physics practitioners, and even then, there was no recognition of the toll of seeing continued deaths of Black individuals by cops. Our experience, understanding and perceptions of events were being manipulated and denied by a racist institution—the silence minimizing the effects this trauma had on Black physicists. Due to systemic inequities, Black Americans are 200% more likely to know someone who's died from COVID, putting us at more risk of declining mental health. Yet even if we are proactive in seeking help, which requires overcoming near-insurmountable barriers, like health care and insurance inequities, as well as stigma, less than 5% of American Psychological Association members are Black. There are growing concerns that White mental health therapists are ill-equipped to provide mental health therapy to Black individuals suffering the effects of institutionalized racism (Torres 2018).

The complete erasure of the effects of the compounded factors of COVID and police brutality on the mental health of Black stakeholders at Fermilab, coupled with a racist, insensitive, and harmful email from the lab, set in motion what we now call *Change—Now*, a strategic call to action by scientists to address these issues and concerns.

As of March 2021, we've dedicated over 100 collective hours of work (not including the parallel efforts, committees, and advisory boards that have been institutionally created) and over nine months meeting an average of two times a week to develop theories of change to re-envision the culture of Fermilab. This is unpaid labor we've given to fix a problem that has been intentionally ignored for decades. There were 5 of us, early career Black scientists at the lab who spearheaded this movement, and soon after we grew to include collaborators who supported the foundational research and development of the calls to action and most importantly, supported us. Our collective has engaged with committees across the lab that we've deemed useful or necessary to continue the path forward to transformational culture change at the lab. We are continuously in communication to support our next strategic move, and we are constantly educating ourselves from the work of change agents, social scientists, and community organizers. Our experience as Black scientists informs many of the actions we take, but as researchers, we also solidify our lived experience with research that sits in Queer Black feminist theory. Our website has a large resource page we used to develop our *strategic plan.*

Change-Now: Vision and Mission of Black Scholars at Fermilab
Vision

We envision an equitable, safe, and just research environment in which Black people can pursue and realize their dreams for scientific discovery. We envision the dismantling of anti-Black racism and other elements of white supremacy that are embedded within structures across Fermilab. Furthermore, we envision the examination and dismantling of the interlocking systems of oppression and privilege within current academic and labor models. The result will be a new community at Fermilab, a new sense of purpose that prioritizes humanity over productivity, accountability over secrecy, and liberation over kyriarchy.

Mission

We will achieve this vision through a people-centered movement that focuses on social change, dismantles systemic oppression at Fermilab, and rebuilds us as a community of individuals with agency and shared leadership. This movement will enable all Black employees at the lab to exercise their collective power to demand transparency of governance and decision-making, with accountability for everyone. Within this frame-work, we prescribe three principles: Recruitment, Representation, and Retention. All three must be implemented to ensure meaningful and lasting improvements. As move-ment-builders, we do not seek to become members of the system that continually shackles the Black community. Rather, we liberate ourselves by seeking to enact our vision through an intersectional lens. Our goal is to increase generational wealth, generational leadership, and build healthy Black communities in and around the laboratory. This cannot be accomplished while prioritizing a focus on compliance and complicity with the current oppressive systems. A complete dismantling of kyriarchies must be embraced by all laboratory constituents and community members.

There is irony in our name "Change—Now" as if by the snap of a finger we can erase the 400 years of generational trauma that we, as the Black community, have faced. We recognize that it will take time, time unwillingly given freely, and time from us doing physics. In Dr Chanda Prescod-Weinstein's book *Disordered Cosmos*, the idea that people perceive time differently is fleshed out. Dr Prescod-Weinstein (2021) describes the notion of colonial time; a perception of time that focuses on capitalism and productivity and that is rooted in white supremacy and the commodification of people. When it comes to scientific endeavors, technology development, and productivity there is inherent time pressure, it must be done now. Colonial time centers productivity over humanity, and commodifies human beings, especially Black bodies, to meet said time pressure. Historically seen in the context of slavery and now in the prison industrial complex and within our academic spaces as we now churn out statements like "Diversity will increase productivity" from the industrial "Equity, Diversity, and Inclusion" mill. Colonial time creates academic cultures focused on perpetually increasing productivity and that was ever so prevalent in 2020 as the main success Fermilab leadership detailed in all-hands meetings was the increased productivity of the lab, during COVID, during civil unrest in our country and a government insurrection, and the deaths of over 500,000 people. This notion of colonial time is both deeply embedded in our academic culture and extremely toxic. This imposed time pressure however evaporates when discussing transformational change, dismantling white supremacist structures, and creating an anti-racist institution. Time then begins to move in slow motion, if it's moving at all. The statement we hear time and time again is "Change takes time." What happened to the fervor we apply to study the cosmos, for looking for Dark Matter, for hunting neutrinos?

In any conflict between technical expediency and human rights, we shall stand firmly on the side of human rights. This stand is taken because of, rather than in spite of, a dedication to science.

—Fermilab's official human rights policy from 1969

In 1968, lab director Robert R. Wilson and deputy director Edwin L. Goldwasser issued the statement above as part of a commitment to the civil rights activism of that era.[5]

Change—Now's goal is for Fermilab to reckon with the notion that, as an institution, they have not stood by their human rights policy of 1969. Fermilab must become accountable for institutional complacency, and recommit to combating economic, social, political, and educational inequities that plague our society. We researched a diversity of tactics that historically have been instrumental to pushing for change in an activists' toolbelt and have been documented heavily by social scientists. Our goal was to use conflict to begin a dialog and imbue a culture shift. We needed to make noise, and the institution's complicity and silence needed to be pointed out.

Since Change—Now's inception, much has changed at Fermilab. The directorate was re-developed to include a chief equity, diversity, and inclusion officer. A diversity and inclusion office was created, separate from human resources, and the staff of this office more than tripled. An equity, diversity, and inclusion task force and senior equity, diversity, and inclusion leadership counsel were erected, with the

---

[5] https://history.fnal.gov/historical/rights/human_rights.html

guiding principles of shared leadership and transparency. A process was developed to submit policy change requests. The importance of community engagement and science communication has been recognized by the institutionalized task codes that are now available to all Fermilab employees; up to 6 hours a week can be billed without the need for prior management approval. Lastly, the importance of transparency and communication has been embedded at the onset of Fermilab's new director's plan as she's spent her first months setting up listening tours of all the groups at the lab, including Change—Now. Dr Lia Merminga has also begun to implement suggestions collated from these listening tours including developing a communication strategy that adds transparency to decisions made at the directorate level and keeps all stakeholders up to date on ever-shifting goals at the lab. We aren't near the culture shift we want to see at Fermilab, but personally, I feel a fervor and drive to create equitable and just spaces for all at Fermilab that I haven't felt in the last five years. I've seen that the creative conflict we developed as a collective has had a positive impact on the future of Fermilab, and I'm hopeful that our continued strategizing and planning will keep the train of progress moving forward. Having a collective live outside of the bounds of the institution, yet adjacent to the institution, has given us more power and flexibility to fast-track many of these initiatives. Working in conjunction with institutionalized processes, we've made incredible progress.

12.4 Black in Physics: Out of the Box

We breathe the same and we bleed the same

But still, we don't see the same

Be thankful we are God-fearing

Because we do not seek revenge

We seek justice

—H.E.R

We are a resilient community. We are continually overcoming and reaching a finish line that keeps being moved. We survive because of community building, because of the support we give and receive, because of the way we use humor to dampen the grief and because we use art and music to help us heal. Querencia—a place where one feels safe, a place from which one's strength of character is drawn, where one feels home—is required for us to thrive. At the end of it all, we are human and it's unbelievable that that even needs to be said.

As individuals, outside of institutions my co-founders and I built a grassroots movement focused on building a strong community of Black physicists with an intersectional lens, to highlight Black scholar's contributions to the field of physics, and to create a counterspace focused on the whole self to counteract the fragmentation we, as Black scientists, deal with on a regular basis. We garnered 2 million impressions, 22,000 profile views, and 2000 new followers in the month of October 2020.

Our Twitter handle has international reach and during Black In Physics week (BIPweek) we had individuals attending our events from places like the UK, Australia, Africa, and Canada. We also commissioned seven pieces written by amazing Black physicists and co-published in Physics Today and Physics World. Having written documentation from Black Physicists in well-circulated and well-respected journals is so important. Not only does it shine a spotlight on the fact that we are here, but it also acts as a historical landmark of the contributions Black scholars have made to the field of physics. I say all this to show how much impact community organizing outside of the traditional institutional box can have.

The most important thing about this movement is that every panelist/speaker/ author and even the Black improv talent we partnered with for our Open Mic night were paid for their labor/their expertise and their talent. If a grassroots movement of 12 organizers starting with no money is able to recognize that the work and expertise of Black physicists and change agents should be valued, have funds raised and worked to make that a reality with no stable infrastructure, there is no reason why well-established institutions should be asking Black scholars to do this labor for free.

We also developed a framework to help train the next generation of leaders in our community. Our organizing team is created from all early career scientists and we built our organizational structure for BIPWeek2021 to have these people in leadership positions, managing volunteers and developing partnerships, community, awareness, and thought-provoking events.

12.5 Key Takeaways

I was recently talking to a journalist; she started listing off some of the movements I'm a part of and ended with "When do you sleep?" It's such an innocuous question with a toxic answer. I've recognized that many of what our field calls "synergistic activities" I do, not because I want to, but because I need to. It took even more time for me to realize that much of the labor I was doing was being taken for free, that I was being tokenized by institutions that were perfectly content with continuing at the status quo. I entered the field of physics to study the weird, and to marvel over the building blocks of the Universe. When I was 8 years old watching the movie *Contact* with Jodi Foster, I imagined my future filled with trying to tackle the world's most complicated topics. I just didn't realize that one of them would be anti-Black racism in academia.

2020 was a tough year. Personally, I lost my father in June 2020. I have also been diagnosed with diabetes, severe acute depression, liver enzyme imbalance, and vitamin D deficiency. I'm human and working at a constant level of survival, knowing the need for structural change for me to thrive in this field and also the suffocating fact that these "extra-curricular" efforts are detrimentally affecting my career, wears on the body and that realization came to a deafening crescendo for me this past year. It's easy to laud the amazing efforts marginalized individuals are making to create a culture shift in your institution, but it's important to note that we didn't choose physics to clean up a mess we didn't create. I'm proud of the work I've been able to accomplish, but the amount of labor I'm outputting isn't sustainable.

Marsha P. Johnson and Sylvia Rivera were trans women of color that were the catalyst behind the Stonewall Riots that sparked the LGBTQ+ movement of 1969. In 1992 Marsha P. Johnson was murdered, her case was originally closed by NYPD as a suicide. Sylvia Rivera was booed off the stage at the 1973 NYC pride fest. These women are the reason I could legally marry my wife, yet they were disrespected in life and in death and are just now being recognized as the leaders they were. I urge organizations not to martyr BIPOC, not to disrespect BIPOC by telling us to our face that you value our expertise and contributions but behind our backs are complicit or silent when we are accosted by white supremacist structures, racism, and discrimination.

I have a duty to speak the truth as I see it and share not just my triumphs, not just the things that felt good, but the pain. The intense, often unmitigated pain. It is important to share how I know survival is survival and not just a walk through the rain.

—Audre Lorde

As Audre Lorde states, this culture shift is not going to be easy, but as BIPOC scientists, we've been embroiled with unmitigated pain and are in survival mode as we maneuver through the halls of these ivory towers. Listen to the truth-tellers at your institutions: experts in social science, and physics education research, center the voices of marginalized peoples, and elevate our lived experience. Carry the burden of striving for change and stand with us, fight with us, stand in the uncomfortable with us. We must think outside of the box, and remove the shackles of the oppressive systems embedded in our institutions. Queering physics means re-examining every facet of knowledge we deem standard or the norm because this knowledge has been curated by a homogenous group that lacks the beauty and color of how BIPOC communities see the world, does their science, and imagines our future. Transformational change takes time, but just like building the next innovative detector and convening the next physics collaboration, we must start now and with the same fervor we have to study the cosmos.

References

Bernard, D. L., Jones, S. C. T., & Volpe, V. V. 2020, J. Black Psychol., 46, 195

Bork, S. J., & Mondisa, J. L. 2022, JEE, 111, 665

Chakraverty, D. 2020, Int. J. Dr. Stud., 15, 329

Isler, J. C., Berryman, N. V., Harriot, A., et al. 2021, Genealogy, 5, 8

Prescod-Weinstein, C. 2021, The Disordered Cosmos: A Journey into Dark Matter, Spacetime, and Dreams Deferred (New York: Bold Type Books) pp 272–273

Robinson, W. H., McGee, E. O., Bentley, L. C., Houston, S. L., & Botchway, P. K. 2016, CSE, 18, 29

Stone, S., Saucer, C., Bailey, M., Garba, R., Hurst, A., Jackson, S. M., Krueger, N., & Cokley, K. 2018, J. Black Psychol., 44, 491

Torres, N. 2018, Research: Having a Black Doctor Led Black Men to Receive More-Effective Care, Harvard Business Review https://hbr.org/2018/08/research-having-a-black-doctor-led-black-men-to-receive-more-effective-care

An Astronomical Inclusion Revolution
Advancing Diversity, Equity, and Inclusion in Professional Astronomy and Astrophysics
Dara Norman, Tim Sacco and Dorian Russell

Chapter 13

Keep Calm and Carry On … Nevermind the Backlash!

Dara Norman

The idea for this book started in 2021, as issues of social justice became part of many conversations in many parts of the US. Science was not exempt from these discussions owing to the obvious deficit of Black, Indigenous and Hispanic/Latinx scientists participating in fields like physics and astronomy, in particular at the higher academic levels when compared to their numbers in the general population. At the same time, astronomy was also undergoing its Decadal Survey Review (see articles by Donahue in Chapter 15 and Bloemhard in Chapter 14). This review process includes input from a wide range of the professional astronomy community and the resulting document is a consensus report outlining the most exciting areas of science likely to be most productive over the next decade. It also provides recommendations for the resources needed to support reaching those science goals. The audience for this report is primarily governmental funding agencies, e.g., NSF, NASA, DOE, lawmakers who authorize and appropriate those funds in both the executive and legislative branches, as well as the community itself that has come together to issue this consensus report.

The 2020 Decadal Survey of Astronomy and Astrophysics (released in 2021, see Donahue in Chapter 15) was unprecedented, albeit not unique, in its handling of issues around workforce and in particular, Diversity, Equity, Inclusion and Access (DEIA). While previous decadal surveys have always, in one way or another, included discussion about the astronomical workforce, the career aspects and the state of the profession, in 2020 the survey chairs codified and prioritized the review of position (white) papers that addressed aspects of DEIA. This was done by convening an official, dedicated panel on the state of the profession and societal impacts—a step that had not been done in the decade before. The panel included sociologists and others who have spent significant time in the promotion of diversity and inclusion in astronomy and physics. The final Decadal Survey report, *"Pathways to Discovery in Astronomy and Astrophysics for the 2020s"* includes

doi:10.1088/2514-3433/ad2174ch13

several high level recommendations regarding improving diversity, equity and inclusion among the astronomy workforce, as well as recommendations for the responsibilities of astronomers to the local communities where their telescopes, and other infrastructure, are located. The straightforward language in these recommendations is truly unprecedented in our field.

In addition to recommendations from the Decadal Survey, funding agencies have begun to incentivize the promotion of DEIA goals as part of the scientific merit of planned research by requiring inclusion plans in which PIs describe how they plan to promote improvements in their research and collaboration cultural environment (Nahm & Watkins 2023). Both NASA and DOE have mandated these "inclusion plans" as a proposal requirement to incentivize principal investigators (PIs) to reflect on and articulate the science they will do and how that work will be accomplished. (For examples, see *NASA's Inclusion plan resources, DoE's Promoting Inclusive and Equitable Research (PIER) Plans.*) Useful tools like the *USELTP's Toolkit of Collaborative Practice* (see Sacco in Chapter 9) are designed to allow PIs to tailor their descriptions of plans to what would be best for their research program and the size of their collaboration, thus not requiring a "one size fits all" approach.

Even before the 2020 Decadal Survey put forth recommendations for advancing DEIA goals, physics and astronomy departments, observatories, collaborations and institutions had already begun the implementation of equitable and inclusive practices. Several descriptions of such activities are captured in this ebook (see Chapters 5, 6 and 7), as well as in several position papers published in the *Bulletin of the American Astronomical Societies' edition of the NASEM white papers on the state of the profession.* Some activities have been spurred on by the release of the AIP TEAM-UP report, The Time is Now (James et al. 2020), a document that outlines steps that departments and universities can take to promote a doubling of the numbers (from 2020) of Black and African Americans receiving Bachelor's degrees in physics and astronomy by 2030. *AIP Implementation Workshops* have been held to help dedicated professors, administrators and students navigate how to turn the TEAM-UP report recommendations into specific actions to further support opportunities for institutional transformation.

Since 2020, institutions have explored new opportunities for recruitment of students and professionals at the conferences of affinity societies like the NSBP, NSHP, SACNAS, etc. Even with the improved recognition among many that biases and barriers to achieving research success exist for many students and researchers, especially those from small institutions or with minoritized backgrounds, those interested in improving the culture in their own local environments have faced stiff "headwinds". Students have described instances where enthusiasm for an improved environment was initially encouraged by faculty, but once a request for the implementation of actual policies, procedures or funding was pushed, that enthusiasm turned into excuses. The inability of some, particularly of those in powerful positions, to recognize or acknowledge these biases and barriers continues to thwart progress. Thus, even in departments and universities where there is interest in improving the climate, gains have been slow. Some faculty have taken the approach that, instead of trying to make too many gains quickly, the community should at least resolve not to do further harm to students as we continue to push for progress. It has been made clear that sustained effort and strategic planning to improve inclusion, diversity, equity, and access in the field are required and significant structural change is likely to take many years to complete.

However, less than 3 years into this awakening of understanding about the missed opportunities for growing our scientific workforce and knowledge base with a highly skilled and highly motivated, previously overlooked source of interested students and professionals, the backlash has already begun in earnest! Recent national politics have led the way to rolling back gains made as far back as the 1960s Civil Rights era. The US Supreme Court decision in *Students for fair admissions, Inc.* vs *President and Fellows of Harvard College* and *University of North Carolina, et al.*, that struck down actions to remedy past systemic and institutionalized barriers in college admissions, has initiated additional concern. Despite the very low numbers of Black, Hispanic/Latinx and Indigenous people in the STEM fields, particularly in physics and astronomy, SCOTUS made the consideration of race in college admissions illegal if used in a "check the box" manner, even when considered only as part of the individualized whole of a student under consideration... except in the case of military academies. However, as discussed below, there are nuances to how an individual student's identity and background can be considered. Engagement with these issues has been so strong in the astronomy community in recent years, that a number of astronomers took part in filing *an amicus brief in support of the case respondents* arguing that (1) diversity is invaluable to ensure the success of the STEM community; (2) that there is data to show that the "Mismatch Theory"[1] is erroneous in its conclusions and (3) that race conscious admissions in STEM education serve as a catalyst to improve institutional culture (Brief for SFFA v. UNC & Harvard College 2023).

Since the SCOTUS decision to overturn affirmative action, there have been a number of laws across the country that attempt to ban broad, nuanced and, in some cases, a more accurate accounting of history around issues of gender, race, and other identities that have been subjected to institutionalized inequities and injustice. Florida state bill 266, limiting the curriculum and instruction in post-secondary educational institutions, potentially imposes a chilling effect on understanding of how institutionalized bias and barriers play a significant role in shaping the makeup of our STEM workforce. As of April 2023, 23 states had introduced or enacted laws that would curb work on diversity, equity, and inclusion. Texas Senate bill 17 bans diversity, equity, and inclusion offices at all public Texas universities, places where faculty could go to learn about best practices for improving inclusive practices in science education. Even more than what these laws actually say or prohibit, the laws are made vague enough to intimidate educators and administrators into self-censorship in order to avoid "getting into trouble".

In the face of these attacks on DEIA principles, how can we continue to make progress toward improving the inclusion of researchers from diverse backgrounds

---

[1] Mismatch theory refers to the suggestion that "students attending institutions where their academic competencies are ranked lower than those typically admitted leads to the mismatched student's loss of academic performance and diminished success later in life because of that underperformance". Concerns about the use of this theory are specified in the amicus brief cited.

and who have had the broad experiences that will allow our field to flourish with new ideas? As a community we need to educate ourselves on what these bills and laws do and do not prohibit. As researchers, we should not be in the business of self-censorship in the face of restrictions that are so detrimental to the advancement and improvement of our ability to do innovative science.

As scientists, we are trained to address complex questions by doing background research, and designing experiments that will provide us with the data needed to answer those questions. In astronomy especially, we design our experiments around the physical parameters of the waves (electromagnetic or gravitational) that we receive from distant objects. Jamie Keith, Distinguished Senior Law and Policy Fellow with the EducationCounsel LLC, advised that the scientific way of thinking about experimentation is exactly the mindset that should be deployed when designing departmental, university or collaboration policies and practices that support diversity, equity, inclusion, and access[2]. We must think of the law not as something to be avoided, but as a parameter that must be incorporated into any design. By understanding how to deal with any limitations that this parameter might impose, we can design around those limitations to achieve our goals.

As astronomers and physicists, we know that if we are designing an instrument, we don't just need to engage those who understand the science to be done. We also need the expertise of those who understand how to construct the instrument. Similarly, if we are to understand current federal and local laws well enough to design for them, we need to engage with those experts to give advice, and the earlier in the designing of our plans, the better. Engaging with those who have the required expertise in the law early in the process is critical to ensuring that we don't start down a path where, at the end, our goals cannot be reached due to an initial design flaw. Consultation early in the process can help to shape our policies and practices so that they are foundationally in line with our DEIA goals. If we wait until there is a fully formed plan to then engage with lawyers or other DEIA experts, we risk having the entirety of our plan nullified as unworkable in the context of the law.

Most universities and institutions employ lawyers and other experts who can advise on the ways in which various aspects of the law might intersect with your plans. When seeking advice, these offices can be good places to get advice on what your institution sees as allowable. It is important to be clear though, that these institutionally retained lawyers are there to protect the university or institution and are not necessarily interested in or have expertise in social justice or civil rights law that may be applicable to what you are trying to accomplish. Therefore, looking for advice from other groups within or outside of your institution may also be helpful to your design plan. Information on good practice for the implementation of DEIA practices and resources can also be found through AAAS' *STEM Equity Acheivement (SEA) Change* Institute. Groups like the *EducationCounsel LLC* provide expert (but NOT individual legal) advice on issues of higher education and the law and have produced a document on *guidance for DEI design parameters under federal law for public and federally funded*

---

[2] (see video, *2023 Diversity and the Law Town Hall*)

private institutions of higher education, societies and other entities. This information is occasionally updated for changes to federal laws.

Professional societies are recognizing that their members need information about these new laws in order to support their students and to do their best research by engaging a wide and diverse community. Information is being shared so that scientists don't have to "go it alone" as they seek to design and implement plans that include making meaningful contributions to DEIA efforts in their collaborations, departments, universities, and in the field. In fact, identifying solutions and activities that have worked in other places may be some of the most useful models. In this book, Chapters 5, 6 and 7 provides some examples of innovative and successful practices that have been undertaken within the astronomy community by collaborations.

Some of these anti-DEI laws, although vague in what constitutes diversity, equity, or inclusion training, are very specific in their focus, for example that DEI offices cannot be set up or that "DEI training" in state funded institutions cannot be required (McGee 2023). However, experience with or training on good practice in mentoring or leadership skills, institutional history, or ethical behavior in science can be a requirement for holding particular leadership positions in a university, department or collaboration in order to effectively qualify for specific jobs. Offering training to faculty who want to advance in their roles within their field could be an effective way to provide some of this content without evoking triggering words.

Although some of the laws prohibit the DEI offices at universities, this may, in fact, allow for proponents of DEI to take a more holistic and integrated approach to many of the issues taken on by those offices. Thus, requiring that a university or department that is interested in the issues of diversity, equity and inclusion not silo work in a single place but instead identify root causes for diversity and inclusion deficiencies and find ways to make solutions part of the culture in which everyone has a stake, could ultimately be beneficial.

Many of the current sets of laws that target DEI offices, do not, currently, target the academic freedom to engage in research on DEI topics. Therefore, inviting speakers who can discuss topics, like stereotype threat, imposter syndrome or equity in STEM, to give talks in academic settings may help students and faculty engage with such topics in a familiar colloquium-style setting. Journal clubs that review academic papers on aspects of DEIA are also good ways to engage the broad community of scientists, and have become staples in several astronomy and physics departments.

In many cases, avoiding the ire of these laws will require collaborations and departments to put in more effort to make their pools of qualified applicants more encompassing of students from widely diverse backgrounds so that a [race, gender, etc] neutral selection can be made.

Neutral programs/policies/selections are important for avoiding confrontation with the law. In this case, neutral refers to those programs/policies/selections that do not, in practice, directly consider any individual's race, ethnicity or gender status when providing opportunities or benefits. Neutral programs can in principle be satisfied by people of all races, ethnicities and genders. Recruitment of targeted groups to participate in neutral programs to satisfy a priority interest in compositional diversity is also possible. One test of a policy's neutrality is whether such programs/policies would be pursued even if they do not also increase the numbers of a minoritized group, although that would be a welcomed outcome. Examples of traditional neutral strategies include increasing access for individuals from low socio-economic backgrounds, enhancing certain geographical diversity, and increasing access for first-generation college students. Although because of structural racism, such criteria may euphemistically select for minoritized groups in practice, they are lawful.

Recruiting students at large conferences or through collaboration partnerships with professors at other institutions has long been part of standard academic networking and has functioned as a way to build careers in the field. As we seek to diversify our collaborations, we should review where and with whom those partnerships are being built. Diversifying and growing our own academic circles is an important way to promote DEIA practice within the field. Leveraging cross-institutional partnerships to engage a more diverse set of students and other researchers can improve both one's science and equity and inclusion goals.

Back in 2020, there was optimism in the fields of physics and astronomy that discussions around equity and inclusion, that should have taken place long ago, were finally happening. There was hope that the stark realities of living in the US as a person of a marginalized identity group were now clear to colleagues who had no reason to consider such concerns previously. However, just three years into opportunities for progress, we are met with a stern resistance to these opportunities for change. It is crucial that we face, and resist, the growing backlash against efforts to make diversity, equity, inclusion and access part of our scientific culture. We must not back down. We have evidence that diverse groups are better at solving complex problems (Hong & Page 2004; Freeman & Huang 2014). Whether we are committed to this fight for moral or practical reasons, or reasons all our own, we will need to stand up to these assaults and continue to work on the plans and strategies that we crafted in earlier times. We need to continue moving forward even as there are those trying to push us back. We will not go back.

References

Brief of Amici Curiae for Students for Fair Admissions, Inc. v. Harvard College and University of North Carolina 2023, https://www.supremecourt.gov/DocketPDF/20/20-1199/232427/20220801151425242_Individual%20Scientists%20Amicus%20Brief%20PDFA.pdf

Freeman, R., & Huang, W. 2014, Natur 513, 305

Hong, L., & Page, S. E. 2004, PNAS, 101, 16389

James, M., & Bertschinger, E.the AIP National Task Force to Elevate African American Representation in Undergraduate Physics & Astronomy 2020, The Time Is Now: Systemic Changes to Increase African Americans with Bachelor's Degrees in Physics and Astronomy (Melville, NY: American Institute of Physics)

McGee, K. K. 2023, Texas Lawmakers Find Consensus on Bill Banning Diversity, Equity and Inclusion Offices in Public Universities (Austin, TX: Texas Tribune) https://www.texastribune.org/about/staff/kate-mcgee/

Nahm, A. L., & Watkins, R. N. 2023, Inclusion plans in Nasa's Science Directorate, 54th Lunar and Planetary Science Conference (LPI Contrib. No. 2806) 1681

Part V

Influencing Science Policy

Introduction to Part V: Influencing Science Policy

In the United States, science priorities and funding are determined through government processes. This part aims to equip astronomers with foundations needed to influence decision makers in government institutions by reviewing the key players, how important decisions get made, and the targeted strategies that astronomers (without policy backgrounds!) can use to collaborate on meaningful policy change at all levels of government.

Advocacy efforts to influence lawmakers throughout democratic systems matter deeply to science outcomes. Whether organized by individuals, professional societies, the National Academies, or other organizations, advocacy influences how the U.S. plans for, funds, and applies science. For many, science advocacy is about more than just great research. Operationalizing equity, diversity, and inclusion (EDI), broadening workforce participation and retention, setting research area priorities, funding universities, developing local and regional research collaborations, and ensuring quality science education are but a few ways engaging with elected officials and advocacy processes can impact the broader astronomy community.

The work of engaging in government processes is *political*—though need not be partisan. In this part, three authors provide insider perspectives on U.S. democratic structures and illustrate where astronomers can be most effective in influencing science policy. We explore opportunities for influence across a wide scale; from persuading Congress on federal research funding (*The Infrastructure Behind National Science Funding and Priority Setting*), impacting generational goal setting by engaging in the ASTRO Decadal process (*The Role of Astronomers in Setting Scientific Priorities through the Decadal Survey Process*). Equally important and often overlooked, this part also explores targeted strategies to impact state and local policies to benefit science (such as dark sky ordinances) in your hometown (*Scientists Belong in State and Local Politics: Strategies to Drive Equitable and Evidence-Informed Decision Making in Government*). By illustrating the political layers of local, regional, state, and national policymaking, we explore areas where astronomers most effectively tap into the process and influence outcomes.

In the United States, stewards of our democratic system have the decision-making power to impact science on a generational scale. Whether elected or appointed, high-level decision makers in our government impact long-range science activity through critical funding allocations, research area prioritization, proposal and compliance requirements, and more. Given this reality, how can astronomers, astrophysicists, and other scientists engage to ensure that their interests are represented in decision-making conversations?

We pose this question at a unique moment. At the time of this book's writing, all levels of government are engaged in high-profile debates on science. Locally elected school boards are revising science curricula; city councils are debating whether to implement pandemic protection measures and consider epidemiological evidence; state-level legislatures are evaluating state budgets and allocations for public universities; and our federal Congress continues to oscillate in support for large scale spending packages in climate resilience, R&D spending, and other priorities. These touchpoints similarly exist for astronomy and astrophysics at all levels of government: for example, local, regional, and tribal governments impact dark sky codes and other facility construction requirements. There are various opportunities throughout the development of these policies for scientists to get involved in the process and help create scientifically informed policies.

An Astronomical Inclusion Revolution
Advancing Diversity, Equity, and Inclusion in Professional Astronomy and Astrophysics
Dara Norman, Tim Sacco and Dorian Russell

Chapter 14

The Infrastructure Behind National Science Funding and Priority Setting

Heather Bloemhard

14.1 Introduction

Elsewhere in this book, the decadal survey process is discussed (see Chapter 15). For those disciplines that utilize a process like this, the decadal surveys play a critical role in policy conversations about science priorities. As informational documents, they are critical building blocks, used by lawmakers, in both science funding and policymaking contexts. Topics that are not covered by the decadal survey process may not have these building blocks, however, they still rely on the same governmental infrastructure for policy making. In this article we will explore the critical components of the federal policymaking process, the key players and the roles they serve, and how these components or players can be influenced.

14.2 Setting Science Policy Priorities

The key players in a policymaking process are, put simply, those people empowered with the authority to make the relevant decisions. If you think the lights in your city are contributing to light pollution, then the key players will be people in your city government. If you're concerned about the state requirements for your school system, then the key players will be people in your state government (see Chapter 16 for more details). In my job, I focus on policy topics that federal policymakers engage in, and therefore the key players that we will be discussing are people in the federal government—this includes people who work in federal agencies and Members of Congress and their staffs.

For most scientific disciplines, including the astronomical sciences, funding is a central concern. Funding for grants to support research and data analysis, lab work, or building instrumentation is essential to the ability of scientists to do their science. Today, the federal government is among the top funders of this work. That funding flows through federal agencies—for the astronomical sciences, the relevant agencies

doi:10.1088/2514-3433/ad2174ch14
14-1

include the National Science Foundation, National Aeronautics and Space Administration, Department of Energy, and the U.S. Air Force.[1] Policies that impose compliance or reporting requirements on recipients of funds or that otherwise impact the ability of agencies to fund research—generally called science policy—can also help, or hinder, the "doing" of science.

14.2.1 Funding Science

Federal spending levels must be determined each year; the White House, federal agencies, and Congress are all involved in the process. The process of setting federal spending levels for science begins with the White House. Following the President's lead on priorities, the White House's Office of Management and Budget (OMB) spearheads the development of the president's budget request (PBR), in close partnership with the relevant federal agencies. Later in this article I will discuss what happens when the PBR is transmitted to Congress. There are a few key steps in the White House process where you may be able to influence the decision makers.

14.2.1.1 The Administration's Role

A president's priorities can be traced back to their campaign. For example, during the 2020 presidential election cycle, the COVID-19 pandemic and the science behind mitigating the impacts featured prominently in campaign promises. However, most candidates don't specifically call out the astronomical sciences during their campaigns. Instead, the astronomical sciences can be inferred from higher-level campaign messages. Science policy experts may be part of the campaign and work to ensure that at least some campaign promises allow for these kinds of inferences. Otherwise, science policy experts will focus their efforts convincing policymakers after they are elected that their science area will support a particular campaign promise. Once a presidential candidate wins their election, they begin to plan for their coming term. The process of translating campaign promises into agency policy begins during the presidential transition. The presidential transition provides an opportunity for agency personnel and science policy experts to communicate the agency's connection to campaign promises with the goal of ensuring beneficial policy decisions are made.

Based on these priorities, OMB works with the Office of Science and Technology Policy (OSTP) to issue an annual memo from OMB to department and agency heads that summarizes the "Multi-Agency Research and Development Priorities." The political leadership at each agency receives OMB's instructions on research and development priorities and works to interpret their agency's role in supporting those priorities. They are usually aided in this exercise by advisory committees, sometimes called FACA committees—which means they must comply with rules set by the Federal Advisory Committee Act (FACA). The Astronomy and Astrophysics

---

[1] Astro2020 was sponsored by NASA, NSF, DOE Office of High Energy Physics, and the Air Force Office of Space Research (AFOSR). https://www.nationalacademies.org/our-work/decadal-survey-on-astronomy-and-astrophysics-2020-astro2020#sectionSponsors.

Advisory Committee, the advisory committees for each division of NASA's Science Mission Directorate, and DOE's High Energy Physics Advisory Panel are all FACA committees. Each includes experts who monitor trends in their discipline and provide advice and recommendations to the agency. One requirement of FACA is that all committee meetings are open to the public, and the agenda often includes an opportunity for public comment. Serving on a FACA committee typically requires researchers be reasonably established in the field, and there is a rigid application and selection process; watching or providing public comment at a FACA committee meeting is not as limited. Either option is a good opportunity for a scientist who wants to ensure their community's opinions are heard by the federal agency.

While the political leadership at a particular agency certainly impacts the overall budget process for that agency, the individuals who handle the day-to-day program activities are critical to the process as well. Federal agencies are staffed by contractors, temporary rotators, or full-time civilian employees, and they often have job titles like program manager, director, or analyst. At research agencies, employees often wish to serve both their science community and the U.S. taxpayer; doing both tasks well requires having and maintaining ties to their science community. Doing so helps program managers understand the needs of their science community and to ensure their agency is a careful steward of U.S. tax dollars. These kinds of jobs are excellent options for any scientist looking to impact their field beyond their specific research area; scientists not interested in completely giving up their research position might be especially interested in serving as a "rotator" through something like the Intergovernmental Personnel Act Mobility Program.

14.2.1.2 Congress's Role

Thanks to Article I, section 9, Clause 7 of the U.S. Constitution, Congress must approve the administration's spending plans before the agencies can execute; more often than not, Congress actually edits and sometimes even rejects the administration's plans for particular budget lines. The annual appropriations process typically starts with the release of the President's Budget Request, then Congress reviews the request and develops their own spending proposals for each federal agency. At its simplest, the process from congressional proposals to enacted federal budget is the same as any bill—for example, reference the 1976 *Schoolhouse Rock!* segment "I'm Just a Bill."[2]

Nominally, it takes Congress seven months—early-February to late-September—to develop the spending plan for the coming fiscal year. However, it actually takes much longer! Congress hasn't approved a budget by the October 1 start of the fiscal year since 1996; the average delay for the past 15 years is around 125 days. The House and Senate Appropriations Committees are primarily responsible for developing their respective chambers' spending plans. Each committee is organized into 12 subcommittees with jurisdiction over a particular set of agencies; most of the

---

[2] https://youtu.be/OgVKvqTItto

federal funding for the astronomical sciences comes from the Commerce-Justice-Science, Energy-Water, and Defense subcommittees.[3]

Around late-February or early-March, each subcommittee holds various public hearings with agency leadership to understand the agency's funding needs in the coming fiscal year. Subcommittee leadership have discretion in deciding which agencies and programs they ask to hear from, and subcommittee members have a unique opportunity to ask agency representatives questions about their programs and policies. Science advocates may try to influence subcommittee leadership when they decide which agencies and programs receive hearings as well as inform subcommittee members' questions. A congressional hearing is not inherently good or bad. A hearing can be a useful way to bring congressional and public attention to the agency, program, or topic, however that attention may provide an opening for criticism.

This timeframe also includes opportunities for other Members of Congress and other community members to provide information about the funding needs of various agencies and is the best time for a person who wishes to influence Congress's funding decisions to weigh in. Science funding advocates can share their views with the Members of Congress who represent them; advocates are hoping that the Members will include particular programs on the lists of "programs important to me" they share with the relevant Appropriations subcommittee. Science advocates may also share their views directly with the Appropriations subcommittees.

After gathering input from various stakeholders, the subcommittees' move the process largely behind closed doors as they consider what funding levels they will recommend. What we know of the process suggests that subcommittee staff "rack and stack" the various lists of priorities they receive from Members of Congress and community members. Our first opportunity to understand how successful our advocacy may have been comes when the subcommittee releases its recommendations ahead of the mark-up and subcommittee vote. At that point, advocates will usually react and share with their Members of Congress what they like and what improvements they hope for.

In a typical process, the subcommittee will advance their bill to the full committee for consideration and mark-up. Advocates will be working to ensure that the parts of the recommendation they like remain in the bill while making the improvements they hope for. At the end of the mark-up, the Appropriations committee will advance their bill to the chamber floor for consideration and possibly further amendment before that chamber eventually votes to approve the bill. Each stage of the legislative process can be much more complicated when you look at the details, but the key take away for advocates is that each opportunity for amendment presents another chance to edit the spending recommendation in their favor.

It is important to notice that the jurisdiction of these subcommittees means that agencies like NSF and NASA are competing for attention with agencies like the FBI

---

[3] The National Center for Science and Engineering Statistics reported in FY 2020 that the top three funders of Astronomy were NASA (70%), NSF (29%), and the U.S. Air Force (0.4%); https://ncses.nsf.gov/pubs/nsf22323.

and the Census Bureau—all are covered by the Commerce-Justice-Science Subcommittee. DOE Office of Science is competing with the rest of DOE and other energy-related agencies; science funded by the U.S. Air Force is competing with the rest of the military budget. Being aware of this can help when you attempt to influence Congress.

14.2.2 Other Science Policy Processes

In addition to setting funding levels, Congress also has oversight powers, which means they review and monitor federal agencies and their policy implementation. This can be as straightforward as asking questions about programs and confirming —or not—appointed leadership. Congressional oversight can also be as complex as adjusting the duties, functions, or structure of existing agencies and even creating new agencies or offices within agencies; these options of influence and control require legislation often called an authorization bill. An authorization bill essentially gives the administration permission—or the authority—to take certain actions; it does not provide the administration funding to do so.

The House and Senate divide oversight responsibility of the federal government into a couple of dozen committees. For NASA and NSF, the House Committee on Science, Space, and Technology (House SST) and the Senate Committee of Commerce, Science, and Transportation have jurisdiction. House SST and the Senate Committee on Energy and Natural Resources have primary jurisdiction over DOE Science, and the House and Senate Armed Services Committees have jurisdiction over the Air Force science programs. Most committees are further divided into subcommittees. Each committee and subcommittee is chaired by a Member from the party in control of that chamber, and the minority party holds the "ranking member" seat. Each committee and subcommittee also has a professional staff whose expertise is more tailored to the subject matter; they support the Members of Congress on the committee and/or subcommittee and their oversight of the relevant agencies.

The Department of Defense (DOD), including the Air Force, is authorized every year as part of the National Defense Authorization Act. The authorization bills for NASA, NSF, and DOE are less frequent, which can limit the opportunities Congress has to exercise its oversight authority of these agencies and, subsequently, advocacy opportunities to influence Congress's oversight of these agencies.

Despite the infrequency of authorization bills, they are an important part of setting science priorities. They can set bold visions for the future of the agency or redirect the agency's attention. They can direct agencies to make changes to how they make funding decisions—for example, directing an agency to fund a certain number of projects in EPSCoR jurisdictions or support projects at emerging research institutions and minority-serving institutions. They can provide an agency with new authorities to support education, industry partnerships, or geographic diversity. All of this impacts how an agency determines their science priorities and the implementation of policies and programs with generational impact.

14.3 Influencing Congress

Weighing in with Congress can take various forms, but, generally, any researcher can weigh in either as a private individual or as part of a group. As a private individual, you can communicate directly with your representative or senators about your priorities; to identify who represents you, please visit: https://www.congress. gov/members/find-your-member.

The ability to communicate directly with your members of Congress is considered a constitutionally protected right; constitutional lawyers and judges have determined that this right applies to groups of people as well as to individuals. An individual scientist can determine which groups to advocate with and then explore which opportunities might be available through that group. Groups that may advocate for the astronomical sciences include your employer, professional membership organizations like the American Astronomical Society, and other non-profit organizations interested in the space sciences. These groups may employ or contract with a lobbyist.

Lobbyists are essentially liaisons between their stakeholders and policymakers. Lobbyists work closely with their stakeholders to know what the most important policy areas are for that group; they monitor the key players for those policy areas and weigh in as appropriate on behalf of their stakeholders. As a result of these relationships, policymakers may come to lobbyists for their subject matter expertise. The stakeholders a lobbyist represents may themselves be subject matter experts, and part of building the relationship with a policymaker is ensuring they know about that expertise as well. For example, if a Member of Congress wants to understand the importance of gravitational waves and how an agency like NSF supports this work, they may start by asking the lobbyist they know; that lobbyist will likely turn to a researcher in the field of gravitational waves and help make the connections. In that situation, the lobbyist may coach the researcher on effective communication strategies, provide the researcher with context for the Member's questions, and generally serve as a guide.

As an individual, you can contact your Members of Congress on any issue; however, you should be aware that most employers have rules related to using work resources to contact Congress. Many employers are required to report the amount of resources they use for lobbying; your communication to Congress while on work time and/or using work resources may need to be reported as well. Federal employees or contractors who support a federal agency are expressly prohibited from using the federal government's resources to lobby because the federal government is not allowed to lobby Congress.

14.3.1 Timing

While there is often a "best" time to share your views with Congress, the average person does not need to stress about timing. It is always a good time to be civically engaged. In general though, the best time to share your views is when your Member of Congress is getting ready to make a decision. For Members of Congress on a relevant committee, that includes in advance of a bill's markup session. For all other

Members of Congress, the best opportunity is when they are about to vote on the bill. The one exception is when all Members of Congress are determining their appropriations priorities in February/March of each year.

14.3.2 Messaging

Once you have decided to weigh in with Congress, and have determined when and how to do that, you'll want to convince your Members of Congress to take some action; what is the process for that? As a first step, revisit your goal. Goal setting experts recommend that goals be based on things that we can control. In this situation, there are many competing factors that influence what your Members of Congress will do, and you do not have control over most of those factors. A more reasonable goal would be to effectively communicate your priorities to your Members of Congress.

Effectively communicating with Members of Congress is very similar to effectively communicating with members of the general public—tell the story of your science, find a way to make it interesting to them, and avoid excessive jargon. A politician's interest can be sparked the same way that any person's can—by capturing their imagination or explaining the importance of your science to their lives. With Members of Congress, you actually have more opportunities to justify the importance of your work, because they also care about the intangible impacts your science might have on the district, state, or nation: Does the instrument or facility you are building support the local economy? Could the science result you anticipate bring prestige to the region? Does your research contribute to the overall competitiveness of the country?

When you communicate your message to a congressional office, you will most likely be speaking with a congressional staffer rather than an elected Member of Congress. Each Member of Congress has staff based in the district or state and in Washington, DC. Congressional staffers are hired by and work for a specific Member of Congress, as such, they will largely reflect their boss's views. However, congressional staff are separate individuals with separate interests and experiences, which gives you an additional audience to consider for your messaging.

Like all science communication, your advocacy message should be tailored to your audience. One benefit of working with a lobbyist is that they will know your policymaking audience and what messages are most effective. Lobbyists can help you understand if Representative A or Senator B will be more interested in the potential for technology transfer or the curiosity-driven elements of your work. They may have met with the specific staffer on this topic and know that they took an astronomy course in college or that they are skeptical on a particular topic. When tailoring your message you might also consider understanding the main employment sectors of the district or state, the committee assignments of the Member of Congress, and what specific action you would like the Member of Congress to take.

Regardless of your specific message, convey it kindly and concisely. If you are sending an email or making a call, be clear about why this matters to you and ask the Member of Congress for something specific. If you are having a meeting, then make that meeting a conversation rather than a lecture. Aim to be honest; congressional staff are particularly good at knowing when someone is misleading them. When talking about your science, be upfront about what it will accomplish or what developments you hope it will support in the future. If you are asked a question you do not know how to answer, congressional staff will appreciate you being honest that you do not know; you can always promise to follow up with more information later—which will help build your relationship with that office. Lastly, remember that you are a constituent of that Member of Congress, and they and their staff have a vested interest in ensuring your interaction is positive. They want you to leave that interaction thinking "my Member of Congress heard me." Even if they disagree with you or are unlikely to do what you have asked, you should expect them to treat you with respect.

14.4 At the Table

There's an adage that pervades the policy community—if you're not at the table, then you're on the table. That translates roughly to the reality that if your and your community's needs are not represented in a conversation then you risk any resources you have being diverted away or otherwise missing out on additional resources. This is meant to serve as a motivation for why advocacy matters, if you want to ensure your community is represented in the relevant policy making conversations, then you need to be involved in the conversation.

Of course, the implied consequence of that adage on the communities who are not represented in policy conversations is that they may be negatively harmed by policies. Our representative democracy is based on the assumption that our representatives will represent all of their constituents' interests. Even for a person who has the mental bandwidth to know everything about every topic, that would be a challenge. For most humans, whose knowledge is based on their experiences and whose lived experience results in particular tendencies and biases, this is an unrealistic task. A completely self-aware policymaker may inherently know when to find subject-matter experts to help them. However, it is more likely that they are not aware of what they do not know until someone tells them about it. In addition to the tangible asks, as mentioned in an earlier section, an important part of advocacy is building relationships with policymakers so they can be a trusted source of information.

If you are already civically engaged—you regularly vote and meet with your policymakers, the next level of ensuring your community is represented could be participating in a FACA committee, working more closely with your organization's lobbyist, engaging in science communication, or possibly navigating a career shift to policy work.

As a lobbyist, I rely on hearing from my stakeholders about what federal policies are impacting their ability to study, teach, and/or research and what that impact is.

A lobbyist may try to gather this information from outside sources, but the opinions of a lobbyist's stakeholders are the gold standard. The pathways for getting your opinions to your organization's lobbyist are too varied to discuss here, but a couple words of advice: be aware of the culture and hierarchy of your organization, understand how your organization makes decisions, and come with questions. Much of the same messaging advice for congressional staffers also applies for communicating with lobbyists.

When you engage in science communication or education you are helping your wider community understand and appreciate science. Future elected politicians, or the staff who support them, will come from that community; if they come into their positions with a well established understanding of the importance of science, then the task of advocating for science will be easier.

If you are inclined toward civil service, you could become a politician (see Chapter 16 for more details). Some local and state government elected positions are part time, so in principle it is possible to keep your research position. For full-time, elected government positions, it is more likely that you might be required to make a major shift in career and lifestyle. A person could shift their career to lobbying, with the goal of eventually representing an organization that allows them to ensure their community is represented, though that can take you further away from actually doing the science that you are so passionate about.

As important as the elected and appointed decision makers are, individual citizens are essential—after all, they are the ones impacted by the decisions. As discussed, being a civically engaged citizen can take many forms. Regardless of how you choose to be civically engaged, remember your values. These create a foundation that will inform your own decision making—is this a topic you feel called to engage on? Is this a situation for you to lead or to follow? Are there potential consequences, if so, are they acceptable to you? Should you push ahead or take time to rest? There's rarely a uniformly "correct" decision, however, building on the foundation of your values and being mindful of the infrastructure of science policy can help as you set out to inform that policy.

An Astronomical Inclusion Revolution
Advancing Diversity, Equity, and Inclusion in Professional Astronomy and Astrophysics
Dara Norman, Tim Sacco and Dorian Russell

Chapter 15

The Role of Astronomers in Setting Scientific Priorities through the Decadal Survey Process

Megan Donahue

15.1 Introduction

The federal funding agencies that support astronomy, NASA, NSF, and the DOE, are Congressionally required to conduct community surveys to inform their scientific priorities. These are the agencies that fund nearly all US astronomical research, with NASA funding about 85% of the federal support going to astronomy.[1] It is difficult to under-estimate the importance of decadal surveys to guide the choices of NASA and NSF over the last 50 years. The surveys guide their strategic planning and their out-year budgets. What's more, these reports are a touchstone for conversation with every funding decision maker in the US government, from the White House Office of Science and Technology Policy (OSTP), the Office of Management and Budget (OMB), to the Congressional staffers who get the bills written and the work done. Recommendations in the latest decadal survey emerge from a community-driven consensus process that is so well-respected, other scientific fields have also adopted single decadal review processes and the emergent reports.

There's never enough money to fund all the projects that astronomers would like to do. The value of a community-wide consensus approach is that it minimizes the bias of individual scientists with outsized political donor connections or other forms of power from having disproportionate sway in putting their projects forward over other agreed-upon priorities. Ideally, the process of choosing how to fund science would be guided by scientific principles. But politicians rarely, if ever, have the breadth of knowledge and experience to understand broad scientific work and its context. So the government has advisory structures and (in the case of many scientific fields today) a cadence of reports that recommend scientific priorities over

---

[1] https://www.aaas.org/programs/r-d-budget-and-policy/astronomy-and-astrophysics; AAAS, Disciplinary Budgets, Joshua Shiode and Joel Parriott, 2015.

time windows longer than the typical term of a congressional representative. Astronomy was one of the first fields to organize and produce a "decadal survey," in 1964. A decadal report with the National Academy of Science's[2] imprimatur and guarantee of process and disclosure of inputs and authors, including bios, etc, is much more compelling and difficult to dismiss than a pitch from an entrepreneurial constituent. A coherent and unified message backed by the community was gratefully welcomed by decision makers, who also realized one must draw resources and people from more than one congressional district, state, or even from more than one country to fund successful science projects.

The astronomy funding agencies turn to NASEM to conduct a survey of scientific priorities to collect scientific input and evaluate it in such a way that financial, personal, or political influences are minimized. The National Academy of Science guides much of the processes for setting up the survey, with input from the funding agencies and from their standing advisory committees such as the Committee for Astronomy and Astrophysics (CAA). The funding agencies are not legally bound by the resulting reports, but they are held to account by both the Executive and Legislative Branch of the US government as well as the scientific communities, which monitor the actual responses of the funding agencies over the years through various external oversight mechanisms (e.g., the AAAC, the CAA, the Space Studies Board[3]) and in the conduction of mid-term reviews (sometimes called the "mid-decadal reviews".)

Astronomers must be involved in this process. The decadal report for a given discipline is accepted by staff and politicians in the Legislative and Executive Branches of government as the scientific community's consensus opinion about the scientific priorities of the next decade.

15.2 A Sea of Communities in a Changing Landscape of Advisory Structures

The National Academy of Science was founded in 1863 by President Lincoln and the National Research Council was organized by the NAS in 1916. The NAS spawned two other National Academies, Engineering in 1964 and Medicine in 1970 (originally the Institute of Medicine).

The National Academies of Science, Engineering, and Medicine (NASEM) was chartered to provide "independent, objective analysis and advice to the nation" and to support science, science education, and scientific communications. It is itself not a government organization, but it reports to the U. S. Congress and it is funded primarily from sponsors, mainly government funding agencies, to pay for various reports (i.e., not just the Decadals). NASEM exists because previous governments recognized the value of science and independent scientific advice to the governance of a country. It is not a coincidence that the NAS was established during the Civil

---

[2] Now the National Academies of Science, Engineering and Medicine (NASEM)
[3] The Astronomy and Astrophysics Advisory Committee (AAAC), Committee for Astronomy and Astrophysics (CAA)

War and that the NRC was established during WWI. To provide the best scientific advice, however, the National Academies are held outside the political process, and many of its processes for studies and their reviews are constructed to identify, and to the extent possible, minimize if not avoid conflicts of interests and the associated undue influence on the scientific advice produced in its reports. Its membership and leadership are elected by its members for distinguished contributions to contributions in their scientific, engineering, and medical fields.

This elected structure, while avoiding the issues associated with political appointees and employees paid directly by the US government, inherently grounds its membership in past structural bias in many ways. Until relatively recently, women, people of color, LGBTQ+ people, and people with disabilities were excluded, sometimes explicitly, from membership.[4]

One does not need to be a member of the National Academy of Science to participate in the Decadal process, but typically the chairs of the decadal surveys have been either elected members of the NAS or that oversight was quickly corrected.[5] It is common to look first to Academy members to fulfill leadership positions. Even with recent efforts to diversify its membership, a typical group of Academy members and panelists as recently as 2010 has been heavily dominated by white men. One would either have to impose term or age limits on members, expand the membership size significantly, or pay explicit attention to who is elected to rapidly change the overall demographics of the NAS membership. However, even as NAS membership updates only slowly, panel membership has been changing as more people are paying attention.[6]

FACA Committees: Federal Advisory Committee Act (FACA; 1972)

At the National Academies of Science, there is an important category of committee called a FACA Committee. FACA Committees are very high-level advisory committees that operate under laws set by Congress in 1972. These advisory committees can be sponsored by federal agencies, and there are fairly strict requirements about meetings, notice of meetings, open meetings (with a few specific exceptions), conflicts of interest and restrictions on members. The process of exactly how FACA committee members are chosen and who does the choosing is often part of a FACA committee's charter. This process, written down, may even be a required feature. Examples of FACA committees include The Astronomy and Astrophysics Advisory Committee (AAAC), the advisory committees for each division of NASA's Science Mission Directorate and DOE's High Energy Physics Advisory Panel.

---

[4] Annie Jump Cannon was denied membership, apparently blocked by Raymond Pearl of The Johns Hopkins University because she had lost nearly all of her hearing. The first woman astronomer elected to the Academy was Margaret Burbidge only by 1978; Vera Rubin was second in 1981.

[5] As in the case of Roger Blandford for the 2010 Decadal survey.

[6] I have noticed a difference even in the last decade. It wasn't long ago that I might be the only woman in the room. I'm happy to report that doesn't happen much any more although we are still out of balance.

In 1997 (Federal Advisory Committee Act Amendments of 1997) the federal government excluded from the FACA "...any committee that is created by the National Academy of Sciences ...". This particular act then applied to the National Academy of Science the rules that it should follow in terms of public notices of meetings, public conflict of interest declarations, and so on. Therefore, there are both FACA and non-FACA committees that give advice to government agencies. The committees established by the NASEM are *not* FACA committees, and thus follow somewhat different guidelines, including that the NASEM does not make it easy to see exactly *how* committee members and chairs are selected since that part of the process is not posted publicly.

Space Studies Science Board (SSB) is an oversight board for the NASEM on the broad topic of space research. It oversees advisory studies, program assessments, facilitates international research coordination, and it also serves as an intermediary between the research community, the federal government, and the interested public. Structurally, the SSB directs the ad-hoc committees. For astronomers, this list of committees include the Committee on Astronomy and Astrophysics (CAA), of which I was a member. But there are other similar ad-hoc[7] committees: e.g., Committee on Astrobiology and Planetary Sciences, the Committee on Biological and Physical Sciences in Space, or the Committee on Solar and Space Physics. Due to overlapping interests, these committees gather annually, typically at the NAS building in Washington, DC, for the Space Science Week. The members of these committees are typically astronomers and engineers working in the space science sector. They may come from universities, consulting firms, government centers and laboratories. These longer-term committees have chairs that report to the SSB, as do the chairs of other ad-hoc committees appointed by the NASEM.

While both CAA and the Astro2020 decadal committees are ad hoc there are significant differences between the two. Until recently, the CAA could not write a report itself. It can now write "concise" reports, particularly to answer limited questions suggested by the funding agencies. The CAA committee was relatively recently chartered to write "concise" assessments of progress against the decadal survey recommendations and to write a concise report on the preparation of future decadal and mid-decadal studies, and to advise the Academies on more in-depth studies to be commissioned, as necessary. There is a lot of activity that gets less press than the decadal surveys but also represents significant community effort. For example, many of these reports generally require a panel of 5–10 people, 2–3 trips to DC and/or Irvine, and weeks of outside labor each to complete. A report is often sent to review, and the response to reviewers is made diligently; each comment gets a specific response and a possible associated edit.

The Committee on Astronomy and Astrophysics is sponsored by both NASA and the National Science Foundation, despite it being an ad-hoc committee under the

---

[7] Note that "ad-hoc" does not mean the same thing to the NAS as it does to most people (even astronomers). To an astronomer, something that is "ad-hoc" might even acquire a pejorative meaning, implying that not much thought or oversight went into the making of it. However, in this case it simply means the committee is temporary and established to consider a specific issue; literally "for this".

Space Studies Science Board. The Department of Energy often also presents to this committee, as the DOE has sponsored, in-partnership with other agencies on other ground based astronomy efforts such as the Vera Rubin Observatory and Dark Energy Spectroscopic Instrument (DESI).

The Decadal Survey on Astronomy and Astrophysics 2020 ("Astro2020") was one such ad hoc study committee. The very first one happened in 1964, and recommended the construction of national observing facilities.[8] Up until then, only a few astronomers had access to large telescopes, telescopes often privately owned or shared between a couple of institutions. The subsequent decadal surveys and other interim studies are listed in this 2019 report diagram, published in 2021.[9]

Some astronomy advice comes from the High Energy Physics Advisory Panel (HEPAP). The NSF controls High Energy Physics's 10 year planning process and report, known as the Particle Physics Project Prioritization Panel (P5). The HEPAP convenes the panels and guarantees the process. In contrast, the Astronomy and Astrophysics decadal review is convened by The Space Studies Board. The CAA drafts the charge and writes the call for white papers, including the prioritization of key topics. The AAAC conducts annual reports to the Secretary of Energy, the NASA Administrator, the NSF director and relevant House and Senate committees to review astronomy activities in general progress with the current decadal survey in astrophysics in mind.

15.3 Service Roles for Astronomers: Committee or Panel Members, Presenters and White Paper Authors

If you are interested in being involved, there are many committee service opportunities to shape the astronomy community's priorities. One route to becoming a member of the SSB or the CAA is to become familiar with the work of the NAS by participating in the various aspects of any of their ad-hoc studies. These studies not only have authors, who meet, collect, receive and digest information for the report, but people who report to those panels and people who serve as "red team" reviewers for the reports. One of my introductory experiences to a National Academy of Science process was a review of Space Data Archives in 2005. Up to that point in my career, I had participated in many NASA proposal and data archive reviews. In 2005, I was about 2 years into my first R1-university academic position as an associate professor at Michigan State University. I was invited as a panelist. Reflecting on the unusual experience of being interviewed by the committee, which was writing a report on astronomical data archives, I felt it would be unlikely I would be asked to return. To my surprise, one of my co-panelists—a professor from the University of Michigan—said something I haven't forgotten, "Well, Megan, once you've been on one of these things they tend to invite you back." Over time I

---

[8] These are the National Optical Astronomy Observatory (NOAO) and the National Radio Astronomy Observatory (NRAO).

[9] https://www.nap.edu/openbook/26073/xhtml/images/img-23-1.jpg

learned that once you opt in for one opportunity, many more will be made available down the line.

I have never been a member of the AAAC but I was a member of the CAA (2012–2019) and several NASEM panels, including a mid-decadal review of the 2000 report (2006–2007), Electromagnetic Observations from Space prioritization panel for Astro 2010 (2009–2010), the NAS AFTA-WFIRST Review (2014), and the Enduring Foundations panel for Astro 2020 (2019–2020). The National Academies did, indeed, invite me back several times, and sometimes (in the case of the CAA service) for multiple terms. I suspect I was invited to serve for many reasons, though what demographic boxes I checked for them were never made transparent. First, I rarely had "irons in the fire", which is a kind way of saying I was not involved in any kind of large project at such a level that the outcome of the study would affect me personally. I was not about to lobby Congress or anybody for multi-million dollar projects. I am also a woman and professor at a large R1 research university and being somewhere in the Midwest, positioned me as underrepresented in this group. I am an observer who uses both ground- and space-based facilities. One could argue I began life a native-speaking X-ray astronomer as a wee undergraduate at MIT, but I came of age speaking theory, UV, optical, and infrared astronomy at the University of Colorado, Carnegie Observatories, and STScI. My training and my inclinations have always been to use whatever telescope best suited the questions I wanted to answer. As a result of being seen as a jack of some trades and a master of none, I perhaps was viewed as an "other" to almost every community. Indeed, on most of the few science-definition-teams I've been on (including more than a few incarnations of the WFIRST/Nancy Roman Space Telescope SDT), I was charged with thinking about the "community science". But that suited the National Academies aims fairly well. My role on any given committee was usually not to look out for my own (if I had any) but to look at the broadest picture.

There are several difficult aspects of committee work. One advantage I had was that my department chair and upper administration at Michigan State University saw my participation in national-level committees as positive, so I could say yes to the call. As a young professor, feeling a little isolated in East Lansing, I decided to say yes to almost everything at first. (I later had to learn how to say "no", because after a while I was being asked to participate on what seemed like every proposal review, and many of those offers being to chair the panel.) Being willing to jump in and review proposals allowed me to meet many people and see how panels and report-preparation worked, first-hand. To the credit of my university and astronomy group, I was supported in doing this type of work.

What I cannot account for properly is the personal (non-monetary) cost of the travel and the absences from my family. I have three children, now grown, but my youngest was born in 2002, so my husband—also an astronomer—and I were often juggling complex schedules. The children are OK today, but life does not make those kinds of guarantees. If I had been a single parent, I would not have been able to do the service work I have done. I don't have an easy solution for that. I have had some reviews pay a stipend for the service, over and above the travel allowance. Some committees—e.g., FACA—are legislated to be purely voluntary, with only reimbursement for travel expenses. There is significant space to evaluate strategies to improve the structure, resources, and compensation available for participants, especially those with family and other responsibilities.

Another hidden cost was to my own science. Undoubtedly, with every panel that I have served on, I have been blessed to learn some new cool science in the process. It is one of the secret reasons I still say yes once in a while. However, the reading and writing that goes into a project review is time I could spend on building a new proposal or writing up a completed analysis. It's a studied trade. Luckily, my university values the national and international scope of my service.

I began serving on advisory committees when I was in my first post-doc as a Carnegie Fellow in Pasadena, California. A casual suggestion by Jeffrey Bennett, (my future textbook co-author), and I was invited to a NASA advisory committee for the Astrophysics Division director. I was invited to stay for a couple more years by his successor. I was a green postdoc at the time, feeling a little out of place at Carnegie and in the rooms of NASA HQ. I sure didn't feel I belonged there with the likes of Riccardo Giacconi, but with encouragement from giants in the field including Andrea Dupree, I eventually grew the encouragement to participate with time. What I learned about how large space missions get finished was scary, but exciting. As a broke postdoc carsharing a used Nissan Sentra with my broke postdoc husband, the notion of influencing multi-billion-dollar research funding decisions seemed foolhardy. However, it was a stone-cold science-inspired maneuver, and it worked, at least in the sense that NASA did eventually get the four observatories launched and two of them are still around, some 30 years after I served on that committee. It is not an understatement to say that Chandra and Hubble dominated my career more completely than perhaps anyone else in the room at that time. Inviting and having young people join in on the big strategic conversations is an investment in the future of astronomy and in their future. Many of the Astro2020 panels included early career astronomers. It's important to have diverse sets of lived experience in the room—e.g., having people who have learned life lessons can balance the irrational exuberance of those who haven't. This split is not necessarily generational, it can be as simple as someone privileged by having experienced nothing but successful proposals, sitting next to me and the rest of us, people who have experienced the far more common experience of mostly rejections.

15.4 Panel Selection, White Papers, Review Process

For Astro2020, an open nomination opportunity was made, and self-nominations were allowed. Once a chair or co-chairs are chosen, they have enormous discretion in how to set up the process. The chairs together with the steering committee of the decadal survey have the responsibility of setting up how the science and the major missions will be prioritized. The agencies, if they so choose, can opt to fund studies. NASA did that for 2020, with significant technology funding and support for four major observatory studies and a suite of "probe" studies, to enable far more reliable analyses of mission costs. However, it is the chairs and the steering committee that decide how to prioritize science and possibly missions.

The 2010 decadal, I believe, was the first to broadly open the door to "white papers" from the community and received over 500. "White papers" are position papers that are not as rigorous as peer-refereed journal or review articles, but they serve the role of pulling together current thinking, summarizing recent work, and collecting input across disciplines including input from social science fields, as in the call for State of Profession papers for the 2010 decadal. In 2019, 573 science white papers were submitted to support the Astro2020 decadal survey, along with 294 white papers describing activities, projects, and state of the profession consideration, the so-called APC white papers. These papers, while limited in length, were sometimes months and even years in the making, and represented a significant investment of time by the community. White papers are often written by large groups but can also be written by a single author.

For the Astro2020 review, each paper, with permission of the authors, was properly published with DOIs in the Bulletin of the American Astronomical Society (Volume 51, Number 3 and Number 11), and the full text of PDFs are available on-line.

Papers were assigned to different panels for review. Some papers were assigned to more than one panel. The papers are thus read by at least everyone in a panel and sometimes more. All the papers are public (if the authors agreed) and available for anyone to read[10] (BAAS Volume 51, Number 3 and Number 7). There were some benefits of being in the room where it (sometimes) happened. For example, while I was on the CAA, I could advocate for paying attention to the timing of the white paper call for Astro2020, the time given to write white papers, and extensions of deadlines. With advocacy from myself and colleague Dara Norman, analyzing author demographics of white papers was considered; however, more culture change work is necessary to convince those in leadership positions of the significant value of this information which could broaden our community understanding of who contributes and potential barriers. In retrospect, author contact information could still allow for a professional demographics survey in the future, even if it only served as a baseline for comparison with future surveys.[11]

I took part in the Astro-2020 process itself, and I was invited to join Enduring Foundations panel. The chairs of the panels have some discretion to choose a diverse panel with the experience and background knowledge needed to review the white papers and presentations for that panel. Reading papers about topics I was more familiar with takes far less time than reading papers about which I have no prior experience—I think that goes for all of us. So even in terms of efficient use of volunteer time, it's a good idea to select people with backgrounds that span

---

[10] See https://baas.aas.org/astro2020-science for science white papers and https://baas.aas.org/astro2020-apc for activities, projects and state of the profession consideration white papers.

[11] By publishing the vast majority of the papers in the BAAS, along with standard information for scientific papers, like email addresses and affiliations of the authors and co-authors, some basic statistics would be relatively easy to gather.

the white papers assigned to that panel. The NAS also keeps track of a set of demographics, representation from a range of institutions, (US) geography, genders, career stage, etc that are all relevant. One opinion I have heard is that it would not be "fair" to invite an early career stage person, but I think that attitude is patronizing. It should be the individual who decides if being on a panel is in their best interest, and for the panel organizer to wonder about how to change the dynamic, so institutions can encourage their early-career researchers to get this type of experience in matters that will be so important for their future careers.

In summary, the white papers not only get read, they get read by a lot of people, from the chairs to the panelists. Sometimes they might be read by the panelists in more than one committee. The white papers present a significant source of information, text, tables, and graphics, that can be cited, borrowed, and integrated into a panel's report.

15.5 Service Roles for Astronomers: Support the Community Consensus

Keep an eye on what the White House (the "administration"), the leadership of the Executive Branch of government, puts in its budget. That is technically binding guidance for NASA and NSF, at least until Congress makes their budget. It helps here to review the difference between authorizations and appropriations. An authorization bill seems pretty important, because it can create a program and in some cases it can provide funds directly. That is called mandatory spending. But all of scientific research funding comes in the category of "discretionary spending". So the authorizing committees make recommendations to the Appropriations Committees which set the funding levels. So Congress can mandate that an agency continue or cancel a program. That means that often the agencies act under direction of law. These actions can take different directions depending on the culture of the agency, but if you witness a civil servant taking a very stressful and/or unpopular decision, recognize they are acting sometimes under a legal mandate to do so. The funding agencies are in the Executive Branch, but their actions can be mandated by the Legislative Branch. That means communications with Congress, and the members of the various committees, can make a difference. Factors of timing, budget phase during the year, what specific bills are under discussion and where they are in the process, where the project is in terms of review and development and whether you are a direct constituent—all of that matters.

The American Astronomical Society (AAS) can help with navigating the twists and turns of the congressional budget cycle and how to advocate on behalf of our collective science goals and in particular the decadal survey recommendations. The AAS has a dedicated policy office that includes a director and a deputy director of public policy, and a John Bahcall Fellow to oversee policy activities at the Society, including support for Congressional Visit Days. The staff provides a bit of coaching, they create some some "leave behinds" (colorful informational flyers) for the

Congressional staff and handle logistics. My experience of talking about science with staffers of all ages has always been positive with the AAS.[12]

The decadal survey is so well respected that a presentation of the Astro 2020 recommendations was made by the Astro2020 co-chairs, to the House Science and Technology SubCommittee (1 Dec 2021). This presentation was a bi-partisan, positive experience. The audience was attentive and the elected officials asked detailed questions about inclusion, Puerto Rico,[13] and outreach to communities in New Mexico. That told me that the inclusion of the very important State of Profession recommendations in Astro2020 is being noticed, and is something that the community can highlight when we communicate with our representatives.

The stewardship of the decadal survey is a crucial role for the astronomical community. We cannot assume someone else is doing all of this. Making sure that the agencies and the lawmakers at least endeavor to follow the recommendations of the scientific community is our task, whether we are working on or with the oversight committees or we are simply paying attention to current events and engaging in community discourse. We can legitimately point out that some of the recommendations are not really about the money, especially in contrast to the sums that they must attend to in the federal budget. They are about the will to do the work, whether it is collecting data, reporting, or creating and enforcing the policies that will make astronomy and the astronomical profession accessible to all of us.

---

[12] On one Congressional Visit Day I got to visit Senator Edward Kennedy's office with John Huchra. John Huchra was the "constituent" and I was the astronomer plus one, but I was thrilled to be sitting in a room with so much informal and (almost) touchable history, including candid photos of all the Kennedys I'd ever heard of growing up, literally covering the walls.

[13] Puerto Rico had been in the news because in 2020, a cable that supported the central platform of the 305-meter telescope snapped, creating a hole in the dish of the Arecibo Observatory. In December the final cable broke, and the central platform fell into the dish. At that time, it was unclear what the NSF's plans were for the Observatory site.

An Astronomical Inclusion Revolution
Advancing Diversity, Equity, and Inclusion in Professional Astronomy and Astrophysics
Dara Norman, Tim Sacco and Dorian Russell

Chapter 16

Scientists Belong in State and Local Politics: Strategies to Drive Equitable and Evidence-Informed Decision Making in Government

Dorian Russell

16.1 Shirking the Myth of Neutrality

Once-upon-a-time pre-2016, I had a hard time convincing PhD scientists and engineers to serve on panels at congressional briefings, join lobbying efforts to increase research and development (R&D) and equity, diversity, and inclusion (EDI) funding, or even to simply sit down with elected officials (aka electeds) at any level of government to share the value of their own research. The pushback I so often received then—and to a shocking degree now—is the misguided concern that STEM professionals cannot engage in advocacy because "science is neutral." This is a great loss to effective governing; this myth lingers despite rampant mis- and disinformation impacting emergencies from climate change to the COVID-19 pandemic as well as years of local, state, and federal assaults on basic scientific principles. While some well-intended scientists indulge in the myth of neutrality, the unfortunate impact of their absence in our political system allows public policies to be crafted without evidence—from flagrant inaccuracies in school-board-developed curricula to state-legislated budget cuts in public health protections.

So-called "neutrality" at the nexus of science and EDI efforts is dually harmful. I have often encountered policymakers and scientists alike who eschew targeted inclusion strategies in favor of "treating everyone equally" in educational, workforce, and other policy areas. This approach denies the reality that our current governing bodies inherited structures that have segregated the benefits and harms of policies both explicitly and implicitly by race and ethnicity, class, ability, and other identifying factors. While many scientists and policymakers will cry out that they do not personally hold the same discriminatory beliefs of governing bodies past, we continue to reinforce and exacerbate disparities by not accounting for generations of policies that segregated benefits.

doi:10.1088/2514-3433/ad2174ch16   16-1

Targeted Universalism (Powell et al. 2019) a concept coined by scholar john powell (who spells his name in lowercase) at the Othering and Belonging Institute, illustrates why "just treat everyone equally" doesn't work. The "universal" part of targeted universalism is aligned with the often non-malicious intention of treating everyone equally: calling for universal goals—the same goals for an entire population. For example, a physics teacher may have a goal that all students in their class show grade-level science proficiency. However, while the goal may apply to everyone, the teacher is likely better off *not* treating everyone the same. For example, if assignments reveal disparities in understanding the content among students who speak English at home and those whose families don't, an appropriate *targeted* strategy may be to offer physics tutoring in students' primary languages so they may gain foundational concepts. Not all students need to access tutoring in Spanish, Somali, etc. Treating groups differently is required in order to get everyone to a universal knowledge level; hence, the "targeted" half of targeted universalism. If the physics teacher decided to treat every student exactly the same, only the students with the privilege of speaking English at home would meet science standards and the goal would not be met. In this example, treating everyone *equally* does not lead to *equity* in outcomes.

16.2 Science is Political—But Doesn't Have to Be Partisan

The students and professionals I mentor, coach, supervise, and champion have all likely heard me joke, "is this a lowercase p or uppercase P situation?" What I intend with this question is to interrogate whether the situation involves partisan (democrat, republican, green, independent, etc) "Politics," or is simply part of government functions and the public affairs surrounding policymaking, i.e., "politics." As we kick off this chapter shaking off the myth of neutrality, we can also relieve ourselves of a burdensome and related myth: that science and politics don't mix. Just because you've dedicated your career to the objective application of the scientific method does not mean that you have withdrawn your rights to contribute to effective government and participate in democratic systems. While science need not be *partisan* ("Political"), defending science and its application in evidence-based policymaking is a *political* process.

Regardless of partisan affiliations, your elected officials represent you, whether you personally voted for them or not. Their time is the people's time—and if you have expertise relevant to important decisions being made impacting entire communities of people for generations, by all means, take their time and make your voice heard!

You can apply your scientific expertise to educate elected officials in order to advance evidence-based policies without engaging in partisan activities. Participating in our democratic system, not in affiliation with a party and/or off the clock if outside of academia, very likely won't get you in hot water. However, there are times where you may calculate that it is not appropriate for you to engage in "capital P" partisan Politics due to employment limitations that could impact professional relationships, or for other personal reasons. For example, some employers limit what activities can occur in the workplace (e.g., don't pass around a sign-up sheet at work for your coworkers to campaign for a Political candidate with you), and some federally-funded positions are not allowed to lobby. When in doubt, check your employer's policies.

16.3 Layers of Opportunity: State, County, City, School District, and Special District Governments

Many scientists are aware that the U.S. lags behind peer countries in the quality of K-12 STEM education; similarly, how many scientists in the U.S. actually received a quality civics education? For our democratic systems to thrive, people need to understand the basic structure and functions of government and how to engage—this is one of many paths toward getting scientists involved in evidence-based policy exploration. Don't worry if you don't recall what you did or did not learn in middle school social studies class. This section is your cheat sheet for determining which elected officials you should engage based on the type of policy you seek to influence.

16.3.1 How Do Universities Get Funded? State Legislatures

Your state legislature (sometimes called a state assembly) is much like our nation's Congress, but localized to your state. Every state except for Nebraska has a two-chamber body akin to the House and Senate with varying nomenclature. While legislative processes, the amount of time the legislature is in session, and other factors differ from state to state, your state legislature makes important decisions on education, infrastructure and transportation, criminal justice, public health and healthcare, and much more. Many astronomers and other scientists are deeply impacted by the funding and operational policies set by state legislatures for colleges and universities, for example. How universities can operationalize equity, diversity, and inclusion strategies, funding levels, and educational and workforce priorities for academic programs and institutions can all be impacted by state legislatures.

16.3.2 Who Impacts Your Local Environment through Things Like Dark Sky Codes and Deciding Where Telescopes Can Be Built? County Commissions and City Councils

Each state is comprised of many county, parish, or borough governments and while they vary widely, all provide on-the-ground services to residents that impact our health, environment, and democratic systems. These governments include front-line services to communities—for example, your county may investigate disease out-breaks, manage elections, and maintain parks where you use your telescopes. Depending on population and geography, your county may provide many residents who do not live in city limits with services typical of many cities (from policing to economic development). Much of the infrastructure related to protecting our health, maintaining democratic systems, and the built environment around us are thanks to counties. While cities may have some areas of service overlap, typically the operation of jails, administration of elections and public health activities are solely counties.

On the other hand, nested below your state and county government, often the closest form of representation you have is your city, town, or village government (if it is not your county). Your city may provide cultural and recreational services, libraries, and manage zoning and building permits (for example, sports arena floodlight locations that may impact your night sky viewing). Your city likely provides emergency response services, manages public works efforts, and environmental stewardship.

Your county commission and city council both are likely single bodies of elected officials, sometimes nonpartisan, with a total number of county commissioners or city councilors you can count on one or two hands. Depending on your county's/city's charter, you may have a board chair or mayor who acts as a chief administrator, or the elected body may select a working professional manager.

16.3.3 Who Decides How Science Is Taught in K-12? School Boards (With Direction from the State)

School boards are special elected bodies with responsibilities only within K-12 education systems. Like county and city boards and councils, your school board is a single body with a number of elected officials you can likely count on one or both hands. Your local elected school board members make critical decisions for your local school district in developing budgets, deciding the policies on how schools run, setting standards for student success, and adopting curriculum. In addition to your school board, your state legislature also impacts K-12 funding and standards. Like with many counties and cities, your school board's elected body selects a professional administrator, called a superintendent, to implement the policy decisions set by the elected officials. Many scientists are keenly interested in the quality of curriculum adopted, how districts prepare students for career and/or higher education success, opportunities for out-of-classroom experiences, and outreach collaborations with higher education institutions.

16.3.4 Special District Boards

In some cases, certain government services directly impacting the public are provided by a special district instead of your city or county. Special districts often have elected boards or commissions with a narrow, focused scope on one specific suite of services. For example, you may have a special district for parks and recreation; certain utilities like water, sewer, and stormwater; library services; fire services. There are countless other examples of service types, often where a regional approach to service delivery is more effective and efficient than if offered by individual cities. For scientists, it is worth diving into special districts; often, these positions can require a great deal of scientific interpretation to function well—whether a soil and water conservation district or internet utility.

16.4 Actions for Residents and Citizens to Directly Impact Local Policy

16.4.1 Voting

Being a citizen does not guarantee your right to vote; for example, previous incarceration status can strip you of that basic right for life in many states. Conversely, some local governments allow non-citizens to vote in local elections only. For example, you may be able to vote for mayor, but not president, so long as you can show residency. If you are eligible to vote, ensure you are registered to vote and participate in every election, not just presidential elections. Do your homework in advance to ensure you are supporting candidates and funding measures aligned with your values; how and whether you vote on every item on the ticket is up to you, the most important thing is you participate in every election.

As detailed in the previous section, lesser-known candidates in offices you might not have even heard of are making the policy decisions that most closely—and swiftly—impact your day-to-day life. Did you participate in selecting the person who is in charge of keeping your drinking water system operational? Who determines how to implement science education goals? Who sets environmental standards like dark sky codes? Who sets strategies influencing what industries and sectors (including higher education) to prioritize in the local economy? The depth and breadth of local government responsibilities is often overlooked and over-shadowed by national debate at a layer of government far removed. Your local governments are your most intimate and most nimble in responding to community needs, and likely have a broader impact on your family than the Oval Office, where federal policies can take years or decades to shift outcomes.

In addition to ensuring you personally vote in all elections, you can have a profound community impact in the spirit of democracy by helping those around you to also vote and develop a healthy view of science and scientists. Within your household, building, or neighborhood block, consider making a "voting plan" each election and speaking one-on-one with folks about items on the ballot impacting science. Consider checking in and ensuring everyone in your household or neighborhood group are registered appropriately (does anyone need a special ballot, such as absentee, for example?). If vote by mail is not available, go the extra mile and contribute to planning scheduled time and transportation to get to the local voting center together. Similarly, use your expertise, voice, and relationships to answer questions and discuss the impacts to science and what's at stake in the upcoming election. Your real-life experience, relationships, and knowledge of the issues can make you an ambassador for science and improve community support for your issues.

Beyond ensuring everyone in your circle is accounted for in your voting plan, identify if you have any time or resources to offer in aiding others. Could your disabled neighbor use a ride to the polls? Can you nominate yourself as snack and water captain and provide for your five best friends while in line at the polls together? If you speak a language other than English, consider the outsized impact you could have by serving as a volunteer in improving voter registration rates for immigrant and first-generation populations. You know your skills and your community needs best: find how they align for a voting plan, including in off-presidential cycles when turnout rates are abysmally low. One vote can be the difference in a local election!

16.4.2 Attending Public Meetings and Providing Public Comment

Like the *Hamilton* Broadway hit by Lin-Manuel Miranda, you want to be "in the room where it happens." Unfortunately for the health of our democracy, public meetings are frequently chambers of empty seats as our elected officials deliberate on major policy decisions impacting university funding, K-12 curriculum, what infrastructure gets built and where, and more. Many boards and councils across the nation have no one show up for public comment or have just a small gaggle of regulars with the privileges of time, high comfort navigating government, and transportation who don't necessarily represent the diversity of community opinions, experiences or even expertise.

There are a diversity of ways to make public comment. Prior to the COVID-19 pandemic, many local government meetings were in-person only, without a virtual video conference attendee option. Now, more and more public meetings are being offered in hybrid format so that residents have the option of speaking from the comfort of their home (or laboratory, or office, or wherever). Whether in person or virtual, most public meetings have a designated time either close to the start or end of the meeting for members of the public to share their views on a particular scheduled topic or any topic of their choosing. In some cases, your local government may require that you sign up to speak the day before the meeting and provide your contact information and the general topic you wish to speak on, so that staff can schedule appropriately and ensure that elected officials can follow up with you if they have questions regarding your comment. Others may have you fill out a note card at the start of the meeting with your information in lieu of pre-registration. Check your local government website for recurring meeting times, one-time meetings on special or emergency topics, and how to sign-up to make a comment.

If you experience any barriers in signing up to comment at a local government meeting—whether a disability, lack of childcare or transportation, or you are just plain shy—you can also provide your views in written form directly to your elected officials. Your local government website will either list an email contact or provide an online submission form for comments. On the flip side, just showing up to local government meetings can be powerful; it is not uncommon for one person to provide comment on behalf of a large group. If the speaking part is not for you, there is the option of helping your neighbors craft your group comment or simply just showing up in person to the meeting in solidarity with similar messaging on signs, clothing, or other creative material in support of the person providing public comment.

16.4.3 Meeting with Elected Officials

While directly speaking to a full elected body during a public meeting can grab the attention of local officials for your cause, you can also make a deep personal impact by meeting with your electeds one-on-one if they are willing and able. Consider calling or emailing your local elected official to request a 15 or 30 minute meeting or coffee chat to discuss your cause. Whether you live in a small town or bustling metropolis, local government is about relationships and can begin to feel much like a small town, despite population size, when you know the few key players. After meeting your electeds a handful of times, don't be surprised if folks remember who you are—or at least what you're asking of them. Building rapport by being available for your elected official to ask questions in a less vulnerable setting—off camera and out of the public meeting hot seat—may be especially helpful if advocating for a science-based policy option that requires deeper background learning.

Depending on the size of the elected body, you may not be able to meet with multiple elected officials at one time, as to do so would reach quorum and constitute a public meeting. It is more likely you will meet one-on-one or in pairs. In preparing to meet with electeds, it is important to know which, if any, serve *at large*, meaning they are elected by and represent all people in the jurisdiction, or whether they are elected in only a specific district or portion of the jurisdiction. If the latter, your voice and interests count the most for the person for whom you are a direct constituent (the person whose district you reside in).

Note that depending on the jurisdiction and role, some elected officials are part-time or volunteer, and may have full-time jobs on top of their elected duties; while they are elected to serve you, consider offering some grace in scheduling. Historically, local governments that have poorly paid or unpaid elected officials have had challenges in recruiting candidates to run for election that are not independently wealthy, retired, childless, or otherwise resourced to take on the labor; this is a core challenge in developing elected bodies that fully mirror the diversity of experiences in the population served.

Questions to ask yourself before meeting with your local elected official:

- What is the problem I am trying to solve?
- How do I know it is a problem?
- Who says it is a problem?
- What is the known opposition, and what evidence do they have for their claims?
- What am I asking this official to do?
- By not taking action, who is harmed? Who benefits from action being taken?
- Are there any potential unintended consequences of my request?

For example, as an astronomer, let's say you have experienced that light pollution in your hometown is impacting nighttime astronomical viewing opportunities for scientists and hobbyists alike. Reading more into the issue, you've also learned light pollution has negative impacts on native wildlife. Digging into your city's website, you learn that there are no ordinances in place relating to dark sky codes or lighting regulations. You decide to raise the issue! To prepare to meet with your city council, you prepare a list of bulleted talking points. You plan to describe your own experience with the problem of not having dark sky codes: you volunteer your time teaching local students how to use beginner telescopes and light pollution has impacted viewing. As a scientist, you read and prepare a jargon-free, elementary-school-level summary of trends you've found in published academic papers in astronomy and the life sciences on the negative impacts of light pollution. You use your training to interrogate sources and spot accurate information. As with any policy issue, you consider how action or inaction impact your community—including how any actions may differ in outcomes by neighborhood, race, ethnicity, gender, class, and other factors. To prepare for potential questions from your elected officials—who very likely are not scientists—you have found multiple credible science advocacy organizations with jargon-free interpretation of academic research to recommend for further reading in a follow up email or printed handout in your meeting.

16.4.4 Serving on an Appointed Committee

Some governments, including cities and counties, have special committees (some-times called boards or commissions) that are comprised of community members, elected officials, or both. Committees are often tasked with making recommenda-tions to the local government on one specific line of service. For example, your county may have a committee that makes recommendations on operating the annual county fair; your city may have a committee that advises on future zoning and planning (can that telescope get built there?); your library special district may have a committee that advises on materials collections and events (how well stocked is the science section?). Often these committees include an application or interview, and you may go through a formal confirmation process with your elected body or mayor. Some committees are quite powerful with requirements set in state law regarding how your elected body uses the committee's recommendations, while others may be purely advisory. For scientists looking to get a taste of local government without running for office, serving on an appointed committee, board, or commission is a great way to have a big impact on services in your community. Scientists may be especially well equipped for these roles, as making policy recommendations can involve digging deeper into data and evidence to make informed proposals for elected officials.

16.4.5 Supporting Campaigns

The first four strategies above largely involve "lowercase p" politics, or advocacy work that need not be partisan. If you have determined party "uppercase P" politics is for you (see previous section), a high-reward opportunity is volunteering for campaigns. Who wins in local elections is often a product of two factors: how many people turn out to cast a ballot, and how many voters were engaged by the candidate. In many local races, whether a candidate wins can be a matter of pure gumption and stamina: how many doors were knocked on, how many phone calls made, how many yard signs raised. Even in high-population areas, local races can be decided by relatively few voters, and at times with paper thin margins. Many voters do not vote in local elections unless they happen to be on the ballot at the same time as major national races (for example, presidential)—but can be convinced if an enthusiastic volunteer speaks to them about the importance of a local race or candidate. Volunteering your time on a local campaign is also an incredible way to connect with voters throughout your community, build relationships, and become more aware of local issues.

16.4.6 Running for Elected Office

Let's be clear: scientists can make great politicians. In an era of science denial, climate inaction, and bungled pandemic response, we could use more scientists in elected office and anywhere decisions are being made. Don't think you're qualified? Behind our current aforementioned policy failures are inexpert decision makers with inadequate understanding of the issues; worse, we have seen unscrupulous behavior from some elected officials. I *promise* your PhD and commitment to improving human knowledge for posterity positions you as more qualified than most of the current slate.

Unlike higher office, you likely don't need to quit your day job to run for school board, city council, and other local positions; likely, you probably can't, as many local elected offices are either volunteer or poorly paid and part-time. If you have the capacity to serve, you're better off considering local office as an additional volunteer role in your life, likely with a four year duration. To get started exploring local office, consider each of the previous five activities above. Through engaging politically, over time, you will build relationships with elected officials and community leaders, understand the landscape of opportunities (vacancies you may wish to run for or incumbents you want to see ousted), and identify supporters and mentors. Volunteering for other people's campaigns is an excellent way to learn if you would wish the same for yourself, and volunteering for committees and engaging current elected officials can help you navigate which kind of roles would best align with your skills and passions. Today, there are also multiple options for local, state, and national programs which offer training for prospective candidates on how to run for office, with many tailored to meet the needs of marginalized people not yet fully represented proportionally in our elected bodies.

Reference

powell, J., Menendian, S., & Ake, W. 2019, Targeted Universalism: Policy & Practice (Berkeley, CA: Othering & Belonging Institute, Univ. of California) belonging.berkeley.edu/targeteduniversalism

www.ingramcontent.com/pod-product-compliance
Lightning Source LLC
Chambersburg PA
CBHW080545220326
41599CB00032B/6370